Competitive Intelligence 2.0

Competitive Intelligence 2.0

Organization, Innovation and Territory

Edited by
Luc Quoniam

iSTE

(W)WILEY

First published 2011 in Great Britain and the United States by ISTE Ltd and John Wiley & Sons, Inc.

ISTE Ltd
27-37 St George's Road
London SW19 4EU
UK

John Wiley & Sons, Inc.
111 River Street
Hoboken, NJ 07030
USA

www.iste.co.uk

www.wiley.com

© ISTE Ltd 2011

Library of Congress Cataloging-in-Publication Data

Competitive intelligence 2.0 : organization, innovation and territory / edited by Luc Quoniam.
    p. cm.
 Includes bibliographical references and index.
 ISBN 978-1-84821-305-0
 1. Business intelligence. 2. Information technology. I. Quoniam, Luc.
 HD38.7.C6576 2011
 658.4'72--dc23
                                                                    2011020211

British Library Cataloguing-in-Publication Data
A CIP record for this book is available from the British Library
ISBN 978-1-84821-305-0

Printed and bound in Great Britain by CPI Antony Rowe, Chippenham and Eastbourne.

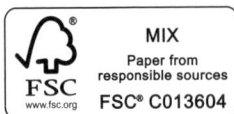

Table of Contents

Foreword

The 21st Century has proclaimed the arrival of a developed world, which is filled with hope and opportunities. However, this century is also full of challenges. Indeed, the information globalization context has influenced the vision of the world in a new way.

This book, Competitive Intelligence 2.0, is a welcome development, as it addresses an elusive field which impresses itself in the economy, social studies, environmental issues, territorial development, innovation, knowledge, etc. Moreover, it is intended to occupy a privileged position as a creativity tool.

Information and world network systems are evolving. From the first generation system, which was predominantly related to the passive storage of information, we are evolving toward a second generation system in which methods, tools, and applications give users new perspectives. This development is generally associated with the suffix 2.0 and testifies to the new capabilities of applications, which should be combined with the countless ways of improving organizational and individual strategies.

For example, technological information processing especially through evolution of bibliometric methods opens new perspectives for innovation. In addition, the progress made in China in translation techniques plays a pivotal role in exchanges between nations, organizations, and individuals.

Thus, the wide variety of topics presented and discussed in this book will offer a broad vision of new aspects of competitive intelligence. The readers will find original articles, as well as relevant bibliographic references, in this book. They will also find various ways of introducing different aspects of Competitive Intelligence 2.0 in their professional practices.

Finally, the decision-maker will be able to develop, without any doubt, the interest in adopting the recommended perception (both sharpened and broadened), allowing us to contemplate the world from the viewpoint of sustainable development and the evolution of values in light of the changing economic models and innovation processes.

This indispensable book should be considered as the domain reference of Competitive Intelligence 2.0.

Zhouying JIN[1]
June 2011

---

1 Pr Zhouying JIN is the holder of the chair of the Beijing Academy of Soft Technology.

Introduction

From the recognition of the disciplinary field of competitive intelligence in the 1990s to the emergence of Competitive Intelligence 2.0, we have witnessed and contributed to the development of tools, practices, and new theoretical approaches based on the current well-established paradigm of "many-to-many" communication. This book is part of the studies conducted by a team, where the team is endowed with 20 years of observations and experiments, from the creation of the first postgraduate degree course dedicated to information (for the benefit of decision-making) in 1989, to the creation of a Research Laboratory 2.0 "Lab4u" in 2009. The "2.0" concept calls thenceforth for the emergence of a new emanticipated Web model with applications in all areas of social activity: management, innovation, education, organization, territorial planning, etc. This book aims to examine the implications of this paradigm shift on competitive intelligence and propose research perspectives.

In the beginning there were semiconductors...

Moore's conjecture in the 1960s described the evolution of semiconductor capability, which doubles each year at a constant cost. Machrone's economic law later indicated that the price of a computer was still, in effect, determined by the consumer irrespective of changes in its technical capabilities. Despite the limitations and contestations witnessed by these economic models, we could, however, observe a continuous increase in the capacity of equipment available to the public for an ever-decreasing cost,

Introduction written by Luc QUONIAM.

which bodes well for a democratization of access to this evolving medium. The promises offered by these technologies were indications of a real change in the world. Graphical interfaces and hypertext navigation brought out tangible changes, and since 1994, each calendar year corresponded to seven technological years thereby receiving the appellation "dog years". This phenomenon is rapidly increasing and a year might correspond to even 10, 20, or 30 years!

This phenomenon can be viewed from several angles; the most simple can be illustrated by the amount of information published annually on the Web, an order of magnitude of several exabytes (10^{18} bits). Considering that a 500-page book represents 1 megabyte (10^6 bits), this quantity is incomparable with the storage capacity of man, which is inferior.

In addition, new features offered by these technologies should be adopted. For example, it is now possible to communicate freely in all languages; this was something unimaginable a few years ago. With language barriers collapsing, competitive intelligence intends to capture knowledge and cultural domains which have been inaccessible so far. Learning these new possibilities and new functioning modes of our society is a complex phenomenon, since it surpasses the updating capabilities of teachers, as well as professionals. The result of this evolution is the emergence of a new "Concept 2.0".

"User generated contents" versus "Publisher contents"

The expression Web 2.0, initially used by Darcy Di Nucci in 1999, refers to the second generation of Web design and development. Tim O'Reilly related this appellation, in 2004, to the idea of cumulative changes in the development of the Internet, as well as its new uses. Compared to 2.0, the early days of the Internet looked like a fake revolution. It was a replication of the traditional information publishing model, where content produced by someone (a "knowledge holder") was imposed and addressed to others, despite the proclaimed possibility that everyone could publish themselves. Therefore, traditional media models (television, radio, etc.) based on vertical communication remained.

A fundamentally different model emerged with about 240 million Websites, in June 2009, and imposed its hegemony. The horizontal communication paradigm, i.e. "many to many" was henceforth dedicated and welcomed with new uses. Two types of phenomena had appeared simultaneously. On the one hand, the vertical communication (one to many) model gave way to a horizontal communication model of "many to many (n to n)" in which a speaker doubles as a facilitator. This is the extension of peer-to-peer functioning in the organization of our societies; network technical architecture transposing into social interactions. On the other hand, this "flattening" of social relations, under the influence of the technical architecture of networks, was accompanied by a community phenomenon.

Applications for collaboration

The emergence of groupware in the 1990s became a starting point for the development of collaboration applications, which allowed users to work "together" and resolve problems of knowledge fragmentation. Web 2.0 is thus regularly called writable Web. This term also expresses the passage from interactivity to interaction through peer-to-peer relationship between users. The peer-to-peer phenomenon thus leads to a method of production, which may sometimes imply the coordination of thousands of volunteer contributors with the selfless goal of completing a project. Therefore the 2.0 tools enable a contributive publishing, which is destined for the collective intelligence of territories and organizations. This aspect shall be developed in the contribution by Serge Chaudy[1] and Lucia Granget (Chapter 13). In fact, these tools would facilitate new information behaviors and communication interactions in everyday life, and more particularly in professional relationships.

These applications described in the contribution by Christophe Deschamps (Chapter 7) can be grouped into categories (VoIP – Instant Messaging Tools, Video Conferencing, Screen-Sharing, Remote-Control, Web Conferencing, Co-browsing, Web Presenting, Work grouping, File Sharing, Document Sharing and Wikis, Collaborative Document Reviewing, Collaborative Event Scheduling, and Mind-Mapping, Project Management). These tools ought to be understood on a broad-scale within the perspective of anthropological changes. The creation of Wikipedia, available for free in 262 different languages, and co-written by thousands of people, is an

---

1 Serge Chaudy is a strategy and communication consultant. His present research works centre on changes in modalities of information production in connection with Web 2.0.

example of the emergence of new publishing models. Despite the dominance of English (about 2 million documents), Wikipedia enables the creation of encyclopedias in other languages, which would not have been possible a few years ago.

These contributive models, which are present in all the socio-economic domains, are accompanied with new information behavior and legal issues in terms of intellectual property and liability (see Arnaud Lucien[2], Chapter 12). In addition, through collaboration tools and the interactive capability of information, social relationships evolve: interaction and contribution create or reinforce social links through a community phenomenon, which is centered on common interests. Blogs and other previously cited applications are concerned, but social networks appear to be the best example of this phenomenon. Different applications are possible: professional networks like *Viadeo* or *LinkedIn* and friendship networks such as *Facebook* or *Myspace*.

The Semantic Web

Aggregation of information and applications from different users and Websites is possible through the use of interoperable languages. This dimension of Web 2.0 constitutes its semantic nature, i.e. the possibility of applications to interact together.

Semantic Web can therefore be defined as a set of interoperable online technologies and applications which interact with each other through a metadata system. These human-generated metadata can be compared to neurons interconnected by synapses. Semantic Web is therefore a form of neural network leading to intelligent Web.

Social bookmarking is a good example of the linking of these elements. This is based on the contribution by Web users who save information resources. The interface allows users to share these resources and rate them, thus adding a possibility to determine their popularity. This information can be consulted based on subject, category, keywords, etc. A list of links is accessible to the public and the links can be made private or shared only among a group of individuals. 2.0 applications network keywords and

2 Arnaud Lucien PhD is a lawyer in information and communication sciences. He specializes in innovation and sociability issues in 2.0 context.

their relationships allow information to be classified, commented on, rated, etc. for innovation strategies. This aspect is dealt with by Brigitte Gay[3] in Chapter 11.

It is also possible to search using tags and generate graphical representations called "cloud tags". There are several applications which allow us to do this. For example, Zotero, a software available for free in more than 30 languages, makes it possible to collect the bibliography that is made available by major information providers (almost all) or by private individuals, and ensures the formatting of the developed document which is totally standard compliant.

From Web 2.0 to "Concept 2.0"

Gradually, the possibilities offered by Web 2.0 have emancipated from the Web to evolve toward Concept 2.0, i.e. a state of mind dedicated to organizations, and based on the values highlighted earlier.

The phenomenon reflects a society where exchanges are more horizontal, networked, and where vertical hierarchy no longer exists. Speeches and organization types are changing, with valorizing individual involvement, and working in ephemeral teams, around various projects. Adaptability and flexibility are the keywords. It is therefore necessary to analyze the relationship between social architecture and shared, collaborative, distributed computer application architectures. "2.0 Architecture" therefore should use a "many-to-many communication (n to n)" in order to ensure collective knowledge. The system should adopt the "permanent beta" status. In the same way, the system and its constituting individuals should continually think out of the box, i.e. integrate a "parallax" as explained in the contribution by Patricia Dupin[4] (Chapter 6) or think "globally" (think global, act local).

We will focus on the keys for implementing a continuous innovation policy. This is why a large portion of this book is focussed on patents.

---

3 Brigitte Gay PhD is Professor of strategy at Toulouse Business School, France. She is interested in driving strategies of alliance networks in complex economic and financial environments.
4 Patricia Dupin is a PhD holder in economic intelligence. She is a specialist in organizational learning and in training on cultural biology.

Technological information, mostly represented by the title deeds of an invention (patents), is discussed in the contribution by Jean Dominique Pierret[5] and Fabrizio Dolphi (Chapter 10). Strategic analysis of patents presented by Henri Dou[6] (Chapter 8) shows the concern for strategic information technology within the reach of developing countries. This aspect can be linked with practices based on "serendipity" and other offensive strategies of industrial property such as those highlighted by Wanise Barroso[7] and Joachim Queyras (Chapter 9).

In fact, the 2.0 concept revolutionizes practices in the innovation field. Rosana Pauluci[8] has demonstrated how a state can be endowed with an information management tool for creating innovation (Chapter 14). These devices are helpful in implementing the "triple helix" model, which is essential for harmonious innovation, even in small enterprises as presented in the contribution by Kira Tarapanoff, José Rincon Ferreira and Lillian Alvares[9] (Chapter 16).

New economic models

One of the changes brought about by the Concept 2.0 and "peer-to-peer communication" is the market evolution (see Chapter 5 by Sébastien Bruyère[10]). The market is thenceforth characterized by the changing demand, management of atypical flows that challenge traditional economic models and organizational conception and management (Miguel R. Trigo[11], João Casqueira Cardoso and Bruno Filipe Carvalho Soares, Chapter 2). At

5 Jean-Dominique Pierret PhD is a specialist in scientific information in the pharmaceutic field. He also specializes in technological watch and in the exploitation of biomedical database.
6 Pr. Henri Dou is a former research director at CNRS. He is Professor of Information and Communication Sciences. He was a former secretary general of ChIN (Chemical Information Network UNESCO), expert for the World Intellectual Property Organization (WIPO).
7 Wanisse Barroso, PhD is a chemical engineer, expert in patent and industrial property.
8 Rosana Pauluci holds a PhD in Competitive Intelligence. Her research works focus on the planning and conduct of prospective studies in strategic sectors for the Brazilian government.
9 Pr. Lillian Alvares' research works center on digital inclusion in micro and small scale enterprises as well as in professional training.
10 Sébastien Bruyère PhD is a specialist in the piloting of projects related to new Web technologies. He also specializes in e-marketing piloting based on Web analytics.
11 Miguel R. Trigo holds a PhD in Information and Communication Sciences. He is a Professor of Competitive Intelligence and Strategic Creativity.

the center of the model is the user, who generates the content that allows an application to succeed. The Internet users are largely encouraged, particularly by the financial profit of the most active and by the media coverage of some success stories, to increase the number of attractive publications, which satisfy their personal interest (ego) and the economic interest of the exploiters.

Arnaud Lucien and Franck Renucci (Chapter 12) show, in addition, that in terms of economic models, the guiding principles of Web 2.0 are gratuity (free) and advertising. In fact, the more the public, the more the profit. The specific 2.0 architecture thus enables working with the global micro-markets characteristics of the long tail. Chris Anderson explains that "those products that are in less demand or record low sales volume can collectively represent market share which is equal or greater than that of the best-sellers, if the distribution channels can propose more choices and create links which facilitate the discovery of such products". In this sense, 2.0 markets are modifying traditional oligopolistic market vision, to return to a more multidirectional, sometimes aggregated, conception, which through the connectivity relationships among economic actors tends toward atomicity. This notion of target market has been clearly demonstrated. Applied at national and territorial (other than geographic) level, competitive intelligence would then become territorial intelligence. Kira Tarapanoff, Lilian Alvarez and José Rincon Ferreira propose an example of territorial intelligence for helping very small firms which are digitally excluded but represent one of the micro-markets of the long tail.

Management 2.0, always connected to Enterprise 2.0, as presented by Miguel Trigo, João Casqueira Cardoso and Bruno Filipe Carvalho Soares (Chapter 6), goes far beyond the use of information and communication technology, even if it is very much present. They base their explanations on market acceleration, requests made to organizations, and evolution of their forms. In fact, organizations are increasingly being called to adapt to volatile markets and meet specific requests in a very short time, while combining risk-taking and high-added value. This acceleration is not unrelated to the evolution of an information society in which dematerialized relationships are eliminating borders and dissimulating hierarchical relationships. This results in new forms of organizations: small pluridisciplinary structures working in high mobility and flexibility conditions. Physically, it is the open space that characterizes this need for communication and project management.

A number of resulting values and principles could be identified: collective goals meeting with individual goals, building of identity within the group and development, at the same time, of strong feelings of belonging, and a permanent link to the team – mobile telephone, chat, Blackberry, connected micro-computers, etc. which may lead to burnout (overwork). In hyperconnectivity, it has become impossible for workers to be "disconnected".

Perspectives of Competitive Intelligence 2.0

As explained in the contribution by Philippe Kislin[12] (Chapter 1), as regards France, "economic intelligence" has emerged as an increasing trend in Western countries and is gaining ground in developing countries (see Henri Dou, Chapter 15 and Amos David[13], Chapter 17), especially through higher education training with necessary cultural adaptations. From the perspective of information system for helping decision-makers, it implies providing good and correctly packaged information at any given time. The term "intelligence" should therefore be understood in the English-speaking sense of intelligence service.

Thenceforth, the 2.0 concept evokes an anthropological paradigm shift in which competitive intelligence involves new tools and a wider scope of action, in so far as competitiveness is being sought in all social activity sectors. Through watch and the constant questioning that it imposes, competitive intelligence is intended to evolve with knowledge and technology. The term "competitive" has to be understood in the sense of constantly seeking improvements in all undertakings geared toward the profit of humanistic values, and far from ambiguities previously maintained in processing "black" information. The contribution by Fabrice Mauléon[14] also establishes a clear link with sustainable development (Chapter 3).

---

12 Philippe Kislin is Associate Professor in Information and Communication Sciences. He specializes in Watch and Economic Intelligence, especially with regard to the translation of decision problems into information problems.

13 Pr. Amos David is the director of studies of a master programme in Technical and Scientific Information – Economic Intelligence (IST-IE) at university of Nancy, France.

14 Fabrice Mauléon PhD is the coordinator and co-author of the book *Management durable: l'essentiel du développement durable appliqué à l'entreprise*, published by Hermes Science in 2005.

The 2.0 is a natural extension for fields of study, as far as the values of the concept and the discipline come together. This latest evolution legitimates, a little more, the compelling need and newness of problematic areas which it is intended to address. Already, Web 3.0 and 4.0 are part of its objects of research in a prospective dimension and predict new practices and anthropological evolutions ensuring sustainability of future researches. This book therefore proposes an overview of Competitive Intelligence 2.0 through the contributions of the authors who by the diversity of their backgrounds, origins, and approaches testify to the richness of this field made open through the "connectivity revolution". In Part 1, we shall consider the impact of 2.0 on organizations, and Part 2 shall focus on the interest of the concept within the framework of innovation strategies. Finally, we shall analyze territorial valorization policies in Part 3.

PART ONE

Organization

Chapter 1

Competitive Intelligence 2.0: A Three-Dimensional Relationship?

1.1. Introduction: From information society boom...

Among the various 20th Century industrial revolutions which have influenced the business world, Information and Communication Technologies (ICT)[1] are part of those that have experienced the maximum turmoil. These evolutions have contributed to the boom in information production and particularly in the demand for information[2], which is essential for all spheres of human activities, for man's adaptation to his environment, as well as for decision-making. From the production perspective, we have observed, as noted by Théry *et al.* "a quantitative increase in information (in its different forms) especially via the Internet, mobile telephony and multimedia production, as well as from other knowledge domains" [THÉ 94]. From the demand perspective, the present consideration

---

Chapter written by Philippe KISLIN.

1 ICT is presently evolving in to a new acronym: ICST (Information and Communication Sciences and Technologies) in order to highlight "their innovating and impacting influence on all business processes either tertiary or industrial" as noted by Nantes economic development agency in an article dated November 22, 2006 (www.nantes-developpement.com).

2 Mavraganis stated it as follows: "Information is no more an intermediate consumption only; it has also become a "vector" i.e. a structuring and energizing force of the productive system that it innervates" [MAV 89].

of "information as raw material" [ROS 96] implies the need to rise up to "this increase" through knowledge specialization.

To cope with this demand for information, new tools, techniques, sharing modes, knowledge exchange methods, and particularly new information search and analysis methodologies have been developed in order to enable:

– acquisition, in the short term and with little or no delay, of relevant documents containing high added-value information which is indispensable for situation clarification and decision-making;

– long-term capitalization and continuous mobilization of knowledge in order to optimize the acquisition;

– storage and protection of information, resulting from situation clarification and decision-making, for future reuse.

In addition, Internet usage and the development of the information highway[3] policy have influenced communication behaviors. Information society, an object of numerous debates[4], has changed the place of organizations in the social landscape and has reinforced environmental complexity. Economically[5], market diversification and globalization have led to increased competitiveness, and organizations in this environment with strong versatility have increased:

---

3 According to J. ATTALI, it is more likely an "information maze", a gigantic tangle of alleys and dead ends, generating distant proximities and deceiving distances. For this author, the Internet would be more like "the labyrinth of a medieval town, with no real architect, instead of an aesthetic organizational highway" [ATT 95].

4 These include the digital advent (and the reduction of its "bill") considered by Curien and Muet as the "third industrial revolution" [CUR 04].

5 Three of the six priority areas of the document "*Préparer l'entrée de la France dans la société de l'information*" (PAGSI) [PRÉ 98] directly concern organizations, and indirectly aim at developing economic intelligence within the following context:
    – information technology, an essential tool for businesses;
    – ambitious cultural policy for new networks;
    – meeting industrial and technological innovation challenges;
    – facilitating the emergence of effective regulation and protective environment for new information networks;
    – new information and communication technology for learning;
    – information technology for the modernization of the public service.

– their environmental scanning capacity in order to discover new opportunities and counter unexpected threats: i.e. watching to stay well-informed;

– their adaptation capacity in responding to changes and new environmental constraints: including promoting sustainable development;

– their reactive ability, so as to be able to rapidly and efficiently redefine their great strategic orientations: this implies engaging in continuous innovation;

– their new ideas for innovation and for remaining competitive: this implies developing creativity.

These capabilities to watch, adapt, innovate, and develop sustainably are what we believe to be the foundations which competitive intelligence is built upon.

1.2. … to the emergence of competitive intelligence

Information society boom and market globalization[6] have opened up new ways of using this concept which has been in existence for a long time[7]. In

---

6 Globalization, here, refers to both the emergence of transnational private actors (with global strategies) and the strengthening of the interdependence of national economic spaces driven by increasing international flows (goods, capital, information, etc.).

7 Many historical examples abound but we shall cite the following:

– the Republic of Venice was able to maintain its power for over two centuries through, on the one hand, a remarkable network of ambassadors across Europe, and on the other hand, a local network of thousands of prostitutes who were questioning the non-residents;

– the Jesuit "monastic" network which, since the inception of the Order of the Society of Jesus by Ignatius of Loyola in the mid-16th Century, have placed "the world in a network" [FUM 05]. Loyola loved characterizing the "Jesuit" (term at the heart of Stendhal's thought) as "a border man", "a man of conversation", who makes a "manifestation of conscience", present in the lively areas, where there are stakes. We will develop these analogies later with the characterization of the role of the watcher;

– many intelligence agencies present in Europe during the same period: the Royal Society in England under Elizabeth I (whose operation was inspired by Francis Bacon and his novel "The New Atlantis"), the "Hanseatic League" of the Netherlands' merchant guilds, the Swedish Wallenberg industrial dynasty, the Spanish "Casa de Oro" or more still the power of Opus Dei, in Germany, the network of General Kaiser during the First World War and in France, the powerful Economic Intelligence of the Pereire brothers creating the "Credit Mobilier" in the mid-19th Century, the Banque transatlantique company, and 17 railroad companies spread over the whole of Europe.

all ages, an enterprising man has always felt the need to be informed, to watch, to defend his "territory", to compare himself with others, i.e. to have the capacity to discern[8], to measure and to evaluate.

The 1994 report of the French General Planning Commission, called the "Martre report", defined Economic Intelligence (EI) as "a set[9] of coordinated actions of search, processing, and dissemination for exploitation of useful information to economic actors, ...these actions are carried out legally with all the necessary protection for the safeguard of the company's patrimony and with the best quality, delay and cost", to the definition proposed by A. Juillet[10], where EI is considered to "involve the mastery, protection, and use of strategic information in order to allow company managers to take the right decision at any given time"; we think that Competitive Intelligence[11] (CI), according to these two definitions, can be seen as both:

– an informational approach comprising a series of operations or processes by which collected information becomes exploitable and interesting. This approach, consisting of "coordinated actions" necessarily involves feedback and adjustments in relation to the decision-making context and knowledge of information needs of economic actors. This approach equally aims at harmonizing the process of search, analysis, dissemination, and protection of information according to the requirements of the context

---

8 According to Michel Foucault, "mind activity will therefore no longer be able to bring things together... but rather to discern i.e. establishing identities and the need to go through all the steps that could alienate from them. In this sense, discerning implies comparing and searching" [LAC 91].

9 This definition only represents, in concrete terms, the first line of a text that gives a more extensive and more complex description of economic intelligence and a dimension of a true social project. This social project refers to the organization and coordination, at the national level, of information exchange behaviors of all stakeholders in the economic development of the nation.

10 A. Juillet was named Senior Official in charge of economic intelligence at the General Secretariat for National Defense in France from 2004 to 2009. This office can be seen as the state's testimony to the importance of EI.

11 In January 2003, French Prime Minister Jean-Pierre Raffarin asked Bernard Carayon to "make an inventory of how France integrates the EI function in its education and training system, public policy and within the business world" and to make recommendations to enhance this feature. The report, entitled "economic intelligence, competitiveness and social cohesion" (The French Documentation, 2003), reflects the work carried out and presents a series of 38 proposals. In order to emphasize the economic and competitive dimension of this report, we refer to economic intelligence and competitive intelligence in a different manner.

and the actors involved. Managing these actions constitute the goal of information watch activity, "infostructure[12]", which is an essential element of the CI approach;

– a mediation between the different actors involved:

- "economic actors" or decision-makers, involved in this process at different levels (from company to the influence strategy of governments or influence strategy of big transnational groups) and often solicited for various roles and functions,

- information watchers (even though they were only allusively referred to in these definitions[13]) responsible for providing relevant information following an informational approach which is explicitly or implicitly formulated. The management of this mediation is the purpose of the collaborative activity between these two principal actors, the decision-maker and the watcher, who are involved in decision-making.

In our own view, ... these two definitions give a leading role to watch and place the watcher (a "pivotal[14]" actor .....and mediation) at the center of our preoccupations. The watcher is thus faced with a two-level problematic issue:

– CI and mediation level: how can he acquire an optimal understanding of the information demand which is formulated from the decision context and inherent associated stakes?

---

12 The neologism "infostructure" (or information structure) according to Combres "allows the development of intangible goods, as infrastructure allows the development of material goods. For the classical highways of industrial society, most of the added value is in infrastructure. For the information society highways, added value and potential for wealth creation are in infostructure" [COM 03].

13 We have extended this study to the 23 definitions found in the appendix of Crayon's report [KIS 07].

14 We refer to the pivotal role, as perceived in group games especially handball. The pivot is the one who plays for others (his team), and must be able to read his team's game, as well as that of the opponent. According to the French Handball Federation's (FHF) lexicon, the center pivot is the player who operates in the defensive system, creating "advantageous space" and serving as support for its entire team. This analogy seems to characterize the role of a watcher as the pivotal actor in CI approach.

– watch and information activity level: how can he "usefully" respond to this demand with the best quality, delay and cost, so that the decision-maker can take the right decision any time?

1.3. CI perceived as a way of managing relationships

It is through this dual aspect, which involves mediation by actions and process harmonization, that we would like to treat CI and consider it as a way of managing relationships between the two main actors (the decision-maker and the watcher) and a third component "information", where the information component could also be called through an extension a third actor[15]. This particular triad allows us to consider three key processes of CI according to the special relationship between these components at a two by two level (Figure 1.1).

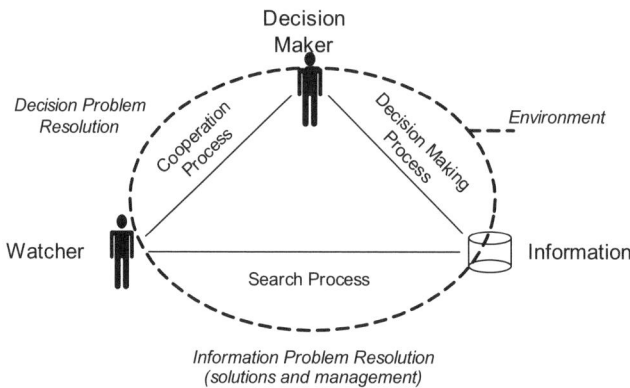

Figure 1.1. *The "three" main components of the CI approach*

There are three processes, where each process joins two components at a time: the cooperation process between the watcher and the decision-maker for collaboratively resolving a decision problem, the watcher information search process for resolving the information problem (which is derived from the decision problem), and the decision-making process between the

---

15 Taken in a very generic sense of the term and by analogy in which information "provides" a much larger role than it does play a role.

decision-maker and the information (made by the decision-maker based on the available information). These three processes are:

– "cooperation" process: this involves the decision-maker and the watcher in a collaboration process where the decision-maker is interested in resolving a decision problem and the watcher is interested in translating the decision problem into information problem;

– "search" process: this is maintained by the preferential relationship between watcher and information for various dimensions of management, mastery, and protection of information;

– "decision-making" process: this process produces a result that emerges from the direct and indirect liaison of the decision-maker and information by filtering diverse mediations.

These specific mediations can be defined symbolically by the "projection" of the remaining components on each opposite side of the triangle. These three mediations will also create an interaction in each of the three processes (Figure 1.2).

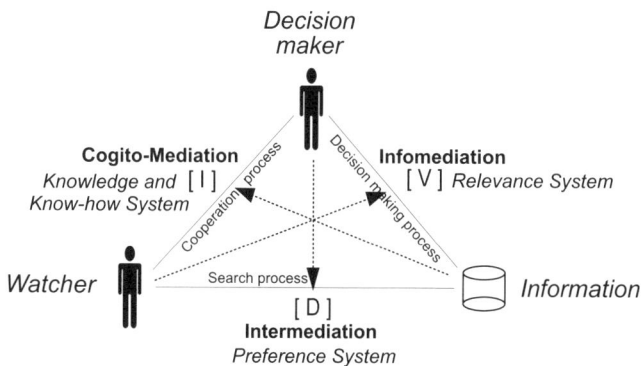

Figure 1.2. *The "three" mediations of CI approach*

These three mediations interact with each of the three processes, and play both the role of a selective filter of preference and that of relevance. They also bring complementary information elements and promote learning through the development of knowledge and know-how. These three mediations are:

– watcher infomediation, which occurs between the decision-maker's formulated demand and the information supply, where the information supply includes the information problem space evaluated in terms of "relevance" (reliability of selected information sources, information processing quality, mastery of domain, time and resource management, readability of documents produced by watcher based on decision-maker expectations);

– decision-maker intermediation, which occurs between the watcher and the information, where the information is conditioned with interferences and is based on adjustments. These interferences and adjustments depend, on the one hand on the decision problem, its nature, the evolutions of its understanding and its context, and on the other hand on the choices made by the decision-maker based on his hierarchical level, his affordance[16], his beliefs, his intuition or cognitive style;

– social mediation of "understanding", which occurs between the decision-maker and the watcher, based on the following two dimensions;

– cognitive dimension, characterized by the understanding of the decision problem stakes and the complementarity of the experience and expertise of the two actors to jointly learn to resolve the decision problem;

– affective dimension, structured around cooperation exchanges between the decision-maker and the watcher, to develop an empathic attitude allowing each one to appropriate the problem of the other. We could join Pennac [PEN 92] to say that this attitude implies that the decision-maker should "gift his preference system to the person he prefers" (here, to the person who acts as the watcher).

We propose that this mediation of providing information in all the cognitive and affective dimensions of learning, cooperation and resolution of decision and information problem be classified as "cogito-mediation" and evaluated "in terms of knowledge and know-how".

We consider the translation of decision problem into information problem in the intermediation, infomediation and cogito-mediation of relationships between

---

16 The "affordances", in reference to the ecological psychology work of Gibson [GIB 77], are literally "indicators that update a range of appropriate and accessible actions in the immediate". We define them as the set of constraints on the decision-maker that determine how he acts and his degree of freedom relative to the situation.

the decision-maker, the watcher and information as the essential feature of the Competitive Intelligence 2.0 process[17]. The valorization of this triple mediation would be the principal success factor of this process since it will ensure trust and consolidate the "co-construction" of exchanges between the decision-maker and the watcher for collaboratively resolving the decision problem. By creating similarity between the decision-maker's preference system and the watcher's relevance system, we can enable the understanding of the decision problem (associated with its stakes and context[18]) and the analysis of information needs to be "proximal" to these two actors.

According to this approach, Competitive Intelligence 2.0 can therefore be seen as a methodology for understanding and resolving the decision problem whose significance is often realized in urgency[19], treated based on the confrontation of two points of view: that of the decision-maker and that of the watcher, and aiming toward a collaborative search for solution.

1.4. Decision-maker – watcher – information triangle: Toward a "bermudization" of actors?

We wish to pay particular attention to various risks which can sometimes come into play within this three-dimensional relationship. Even though a triangular form often tends to reflect some stability[20], it is also used as a

---

17 The 2.0 concept in our context does not imply a version number, like in software, but a shift from information-centric vision (we believe the goal of Web 1.0 was to recreate a world library) to user-centric vision. By analogy to what Tim Berner-Lee called Web 2.0, Competitive Intelligence 2.0 is an actor-centric activity, whose purpose is to give "value to meaning" by making actors with different skills collaborate.

18 The understanding of the information demand associated with stakes and decision problem context is what we called DSC [KIS 03].

19 The socio-economic environment of organizations is characterized by rapidly increasing product renewal and upgrade cycles while demanding a necessary effectiveness of committed resources. This strong movement makes the decision-maker often "act in urgency and decide in a situation of uncertainty" [PER 99] (expression borrowed from P. Perrenoud in analogy to his definition of teaching).

20 The triangle is a symbol of stability: as in pyramids, roofing profiles, civil security (blue triangle in an orange circle). It is also used as a solid foundation base as in the French Republican triangle (liberty, equality, fraternity) and in the trinity of the Christian religion. "The stability triangle", well-known to forklift designers, is also present in the balance of any construction element.

symbol in chemistry[21]. It is sometimes used to represent impossibilities[22] or danger[23]. Drawing inspiration from the work of Houssaye[24] [HOU 88, HOU 93] and transposing this into the context of our actors and components, the major equilibrium risk of our triangle is that, two of these actors are inclined to maintain a special relationship (and recognizing each other as partners) by tending to exclude, in certain situations, the third actor who by "bermudization" would be constrained to death place or by default play as the crazy (Figure 1.3).

The risks are concurrent to the three processes (search, cooperate and decide). Since it is difficult to consider the three axes as equally important, two of the three components could be accorded almost equal importance. This would force redefining of the third component based on the other two components. This redefinition can be tends to break or craziness (the watcher can terminate his contract of informing the decision-maker by objecting to his authority or breaking the relationship) or withdrawal or

---

21 The triangle is associated with the father of modern chemistry, Lavoisier (1743-1794) who was the first to demonstrate that water is the product of the combustion of oxygen and hydrogen. The mechanism of combustion is called the "triangle of fire" because three elements are necessary for its existence: fuel, oxidizer and energy. Removing one of these three elements leads to its immediate extinction. We also find the triangle in alchemy, where it represents sulfur (a triangle surmounted by a cross), which gives it an occultic and sometimes evil connotation.

22 Like the Kanizsa triangle (optical illusion), the Penrose triangle (impossible object designed by Penrose in 1958 and used by Escher in his paintings), or the love triangle (husband, wife, lover) of Tartuffe and of the Vaudeville theater, is the source of many misunderstandings. We also find this element of impossibility in "Mundell's triangle", used in economics and finance domains. Mundell's triangle represents the three principles for judging the viability of the international monetary system (IMS): degree of stringency of exchange rate, degree of capital movement and degree of monetary policy autonomy. Mundell showed that it was impossible to reconcile the three vertices of the triangle at the same time. For example, when the capital moves freely and the exchange rate is flexible, a country can no longer freely set her interest rates, and her economic policy is therefore dictated by the movements of exchange and not by the public authorities of the country.

23 The triangle is used to signal danger (in road and fire safety codes). In sociology, we find the Karpman triangle or "infernal triangle", which shows that individuals can be locked up in a cyclical manner in the position of victim, persecutor or rescuer. Various forms of danger exist in several variants of the triangular interaction like in any paradoxical situation, rich in interaction, where the opponent is also a potential partner.

24 J. Houssaye highlighted the interactions between knowledge, master and student (the three poles of the didactic triangle in the act of teaching).

death (a decision-maker may defer on several occasions the decision-making even though he is in possession of all the required information elements).

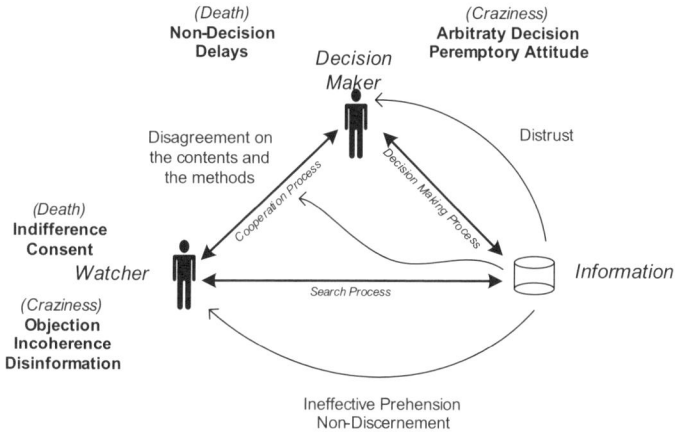

Figure 1.3. *The different risks of actors' "bermudization" within the decision-maker-watcher-information triangle*

A partner, in this context, is an actor or a component (if it is information) with whom it is possible to maintain a preferential relationship. A partner is therefore one, or what allows one, or more of these elements to exist reciprocally or preferentially.

The dead person is the one that places a gap in the relationship and who, in consequence, is no longer recognized as part of the relationship. Its mode of presence is more of absence than of real reciprocity. By analogy, the dead person would be the character in the bridge game who plays a minor role but is paradoxically indispensable since it is not possible to hide him by making him play more than he plays. Its position could have been assigned, defined and conducted by others who would be the real actors of the situation.

The crazy person is the one who objects to the rules and relationship with respect to the excluded one, and is no longer allowed to establish a contact, affirming his negation by a speech or a refusal, and creating some hardly controllable situations.

For each of the axes of our triangle, we would have:

– the "search" process would consider the special relationship between the watcher and the information. Here the risk of death or craziness of the decision-maker could delay decision-making. Alternatively, the decision-maker may not take any decision with respect to the information received, or would tend to play the crazy one by displaying some authoritative attitudes or taking some arbitrary decisions, i.e. by neglecting or not considering the product to be watched;

– the "decision-making" process is the result of the direct and indirect liaison between the decision-maker and information, where the risk of death or craziness with regard to the watcher would persist as a result of the indifference or consent of the decision-maker. This could manifest in the objection of the watcher to the cooperation by providing incoherent or non-relevant information, even to the extent of mis-informing or dis-informing the decision-maker;

– the "cooperation" process is characterized by the non-definition in advance of the rule that guides the decision-maker-watcher relationship. In this process, the actors will have to jointly define and constitute the rules in order to specify the way they will integrate and transform the problems into solution i.e. transforming information into decision. How do we imagine the making of the information component to be "the dead one" or "the crazy one"? Quite simply, when information can no longer be "controlled" and it strongly influences the "acts" of the other two actors as well as the processes. We can thus find an attitude of deep distrust on the side of the decision-maker, an inefficient informational perception or non-discernment on the part of the watcher in his information search activities, or even disagreement on the contents and methods within the cooperation.

This triangle registers in other triangles and is influenced by them. Among these influencing triangles, particularly at the enterprise level, we have: the strategic triangle [TAR 91] which highlights the interactions and the mutual dependency between strategy and enterprise, its structure and information technologies; the management control triangle[25] [ANT 88]; or

---

25 The management control triangle connects the resources allocated to the decision-maker, the objectives set for him and the accomplishments. Relevance is then defined as the relationship between goals and means devoted to them. Management control, according to Anthony, is "the process by which managers influence other members of the organization to implement the organization's strategies" [ANT 93, p.10].

by adopting a systemic approach we find that the Mélèse triangular representation [MEL 92] combines control system, management system and information system. In addition, from the socio-emotional angle, the actors find themselves under the reciprocal influence of their mood and personalities, which implies that each of them has to find a good relational distance within the collaboration.

Apart from the different cognitive biases which exist within the triangle, the one that appears more important for us is the "loss of meaning" (a somehow "Bermuda" effect). Loss of meaning plays an important role in the collective validation of the product of watch or all decisional elements, since each participating actor is allowed to give his or her interpretation of the product. Loss of meaning thus creates ambiguity since it opens the door to all possible interpretations by making all validation methods acceptable, irrespective of the initiated processes. Another effect of the loss of meaning is the "too immediate" passage to solution because it is easier to produce a solution than to ponder on the objectives. In other words, as highlighted by Morel, "it is much easier to engage in technical activity than in political activity" [MOR 02]. From the cognitive point of view, developing an objective requires broad vision, abstract concept handling and considering several alternatives. It is thus easier for a watcher to develop an information retrieval tool than to define the usage objective of the tool.

To minimize these conflicts and bias, it is necessary that the actors develop a cognitive dialog and explanation i.e. the social sharing of a common "meaning store" with some stability and objectivity. This implies that fostering collaboration and the performance of our triangle depends on the existence of a truth standard and common understanding.

1.5. Teaching companies to be "intelligent": competitive versus competition?

The novelty of the concept of CI with the 2.0 flavor is the realization that companies possess knowledge and know-hows (and more importantly "collaboration know-hows") for resolving decision problems which cannot be limited to individual competence. This knowledge, certainly, is transferable from individual to individual, as well as from collective to individual and from collective to collective. It can therefore be considered as a collective interpretation system, capable of learning and "in which the members can continually expand their capacity to achieve their desired results, where new thinking modes are developed, where collective aspirations are not hindered, and where people are continually learning how to learn together" [SEN 02]. We can therefore talk about an "intelligent" company in the same vein as Varela who also added that "it is not the problem solving skills of an organization that makes it intelligent, but the skill of its members to create "a world of shared understanding", a cognitive act which involves listening to one's colleague and welcoming everyone's unique perspective" [VAR 93].

However, the principal obstacles to this Intelligence 2.0 are also the principal obstacles to sustainable changes in organizations: "too" linear thinking mode, competition valuing, and tendency to react rather than engaging in collective thinking. Kofman and Senge [KOF 95] attribute most of these recurrent difficulties to three "ancestral" human attitudes: fragmentation, reaction and competition. These attitudes influence the processing of a decision problem within our triangle:

– fragmentation is a natural tendency in human beings to solve a seemingly complex problem by initially decomposing the problem into seemingly "simple" parts and thereafter studying each part separately. In an organization, fragmentation can manifest through departmental partitioning, which contrast with one another (or make them ignore one another) once a problem surfaces or makes them to have a reflex to quickly search for the guilty, even before bringing together the competences that could solve the problem;

– reaction is related to the fact that, in most of the period in human history, man had always faced sudden threats such as wild animals, flood, earthquakes and tribal attacks. His nervous system has evolved based on these conditions. That is why man is more "prepared" to react to sudden aggressions than slowly evolving aggressions[26] e.g. deterioration of his environment. This tendency to react spontaneously as soon as a problem occurs, could make the decision-maker prefer a solution approach (by trying

---

26 Peter Senge illustrates this phenomenon by using the "frog" metaphor [SEN 02]. If a frog is immersed in boiling water, the heat will force the frog to try to jump out of the pan and save its life. However, if it is immersed in a pan of cold water and the water is gradually warmed, the frog will die scalded, because the slow and gradual evolution of the temperature does not allow it to define a tolerance and action threshold. This metaphor is often used to show the importance of "timely" perception of weak signals emerging from within and outside of an organization and proclaiming the need for implementing major changes.

to "quickly" get rid of this "bad situation") that he knows, rather than a more creative approach which demands interactions and a more profound reflexion time;

– competition, according to Mathieu, is the worst enemy of learning: "to learn, we must realize that there are some things that we don't know and we must also try some activities that we have not "mastered" and allow others to help us. But in most of the organizations, confessing one's ignorance is always considered a weakness; our value depends on what we know and not on what we are learning" [MAT 03]. Another adverse effect of competition is that it makes the decision-maker more focused on short-term measurable objectives. Thus, when a decision problem surfaces, he could, under the influence of competition, be forced to act quickly, rather than take enough time to ponder (which implies confessing that he does not have an immediate solution). The decision-maker may even adopt drastic measures (such as down-sizing, general "rationalization" of cost) while remaining in doubt as per the possibility of these steps attacking the underlying problem. This logic of "quick fix" will have more devastating effect on the understanding of the whole problem.

1.6. Conclusion

Throughout this chapter, we have elaborated different relationships, mediations and processes within a triangle formed by the decision-maker, the watcher and information. This triangle is a transposition of the well known teaching triangle in educational sciences made up of master, knowledge and student. Another analogy with the teaching world can be highlighted. A teacher (whose definition originates from ancient Greek (*paidos* and *agein*)) is the one who takes a child through the city for exploration and discovery of meaning. The watcher as a teacher is the guide. He guides the decision-maker in resolving a decision problem and helps him to explore the environment by providing him with meaning. The first person is a mediator between knowledge and the child while the second is a mediator between information and the decision-maker. One develops economy of intelligence and the other develops economic intelligence, and the two provide value to the meaning. They are both convinced that it is necessary to listen in order to be heard, and that their relationship is built on mutual trust [KIS 09a].

The goal of this mutual trust is to facilitate the translation of a decision problem into an information problem. To establish this translation, we propose acting on all three processes by optimizing the three mediations which characterize them (Figure 1.4). So, creating a communication interface between the decision-maker and the watcher will prevent the two actors from inventing missing information: for one, probabilities of making delusive choice, and for the other, proposition of inadequate information content and form. Similarly, an assisted description of the decision problem context from the emergence of its stakes, associated with the initial demand of the decision-maker and its different formulation and reformulation by the watcher, gives a possibility of increasing understanding of the decision problem and its consequent translation into an information problem. Finally, the follow-up and supervision of search will be possible due to storage, annotation and archiving of all information elements in order to facilitate "cognitive traceability" [KIS 09b] for possible future reuse.

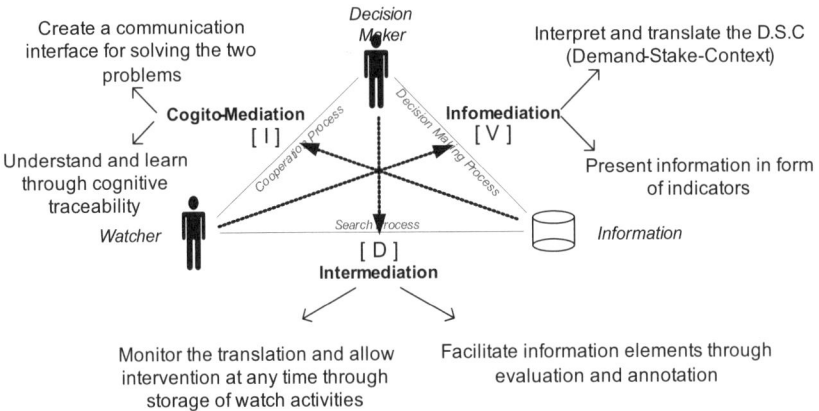

Figure 1.4. *Facilitate the entire three processes*

Two main actions are taken in each of the processes: a "material" oriented action (a communication interface (software), presentation of information as indicators, storing of search activities) and an "intellectual" action (or non-material such as understanding of the Demand-Stake-Context (DSC), supervision of the entire process and the reuse by cognitive traceability).

We will close this chapter by highlighting the apparent contradiction between the economic theory which advocates that the value of a good depends on its scarcity and the definition of information whose value increases in proportion to its circulation. As noted by Amidon, "the economic theory has a problem with knowledge: knowledge defies the scarcity principle; the more it is shared, the more it is developed; the difficulty is in proper management of this precious resource" [AMI 01]. This divergence as noted by Fayard [FAY 05] is like Oscar Wilde's maxim: "to the economists, everything has a price but nothing has value".

Conversely, having scarce and "secret" information can be a competitive advantage. The value of this information will therefore be very high, and would not be based on its circulation, because it will never be exchanged, but based on how the decision-maker decides to use it. However, for the watcher, the real value of information is a prospective value: that of the quality of the relationship he hopes to maintain with the information. Enjoying a good relationship with information while being faithful to the relationship with the decision-maker will be the stake and the goal of a true three-dimensional relationship in the 2.0 era...

1.7. Bibliography

[AMI 01] AMIDON D.M., *Innovation et management des connaissances*, Editions d'Organisation, Paris, 2001.

[ANT 88] ANTHONY R.N., *The Management Control Function*, The Harvard Business School Press, Boston, 1988.

[ATT 95] ATTALI J., "Les labyrinthes de l'information", *Le Monde*, 9 November 1995.

[COM 03] COMBRES C., Réseau : L'infostructure, une solution pour les hauts débits métropolitains, 2003, www.afnet.fr (accessed in March 2009).

[CUR 04] CURIEN N., MUET P.A., "La société de l'information", *Les rapports du Conseil d'analyse économique*, no. 47, p. 310, La Documentation Française, Paris, 2004.

[FAY 05] FAYARD L., Techniques de communication, Maîtriser son information, Notes de cours, UFR Master 2 économie TQ2E, University of Paris 1 Sorbonne, http://lucfayard.blogs.com (accessed in July 2009).

[FUM 05] FUMEY G., BOËDEC F., "Comment les Jésuites ont mis le monde en réseau depuis le 16e siècle ?", in *Le monde en réseaux. Lieux visibles, liens invisibles, 16e Festival International de Géographie*, Saint-Dié des Vosges, France, October 2005.

[GIB 77] GIBSON J.J., "The theory of affordances", in SHAW R.E., BRAD-SHAW J. (eds), *Perceiving, Acting and Knowing*, Lawrence Erlbaum, Hillsdale, New Jersey, 1977.

[HOU 88] HOUSSAYE J., *Le triangle pédagogique*, Peter Lang, Berne, 1988.

[HOU 93] HOUSSAYE J., *La pédagogie : une encyclopédie pour aujourd'hui*, ESF, Paris, 1993.

[KIS 03] KISLIN P., DAVID A., "De la caractérisation de l'espace-problème décisionnel à l'élaboration des éléments de solution en recherche d'information dans un contexte d'Intelligence Economique : le modèle WISP", *IERA'2003*, INIST, Nancy, 14-15 April 2003.

[KIS 07] KISLIN P., Modélisation du problème informationnel du veilleur dans la démarche d'Intelligence Economique, Thesis, University of Nancy, 2 November 2007.

[KIS 09a] KISLIN P., "Tracer, Annoter et Mémoriser : Trois Actions pour Asseoir la Collaboration-Confiance du Veilleur et du Décideur", *Actes de VSST'2009*, Nancy, 2009.

[KIS 09b] KISLIN P., "La traçabilité cognitive : fil d'Ariane du veilleur pour aider le décideur à sortir du labyrinthe décisionnel", *Intelligence collective et organisation des connaissances, Actes du 7ᵉ colloque international du chapitre français de l'ISKO'2009*, University of Lyon III, June 2009.

[KOF 95] KOFMAN F., SENGE P., "Communities of commitment : The heart of learning organizations, organizational dynamics", *Learning Organizations, Developing Cultures for Tomorrow's Workplace*, vol. 22, no. 2, p. 15-43, 1993, 1995.

[LAC 91] LACOUTURE J., *Jésuites 1. Les conquérants*, Le Seuil, Paris, 1991.

[MAT 03] MATHIEU A., "Les organisations apprenantes et les défusions", *L'Agora*, 10, 1, été 2003.

[MAV 89] MAVRAGANIS N., "Information, culture et société : la montée des réseaux", *Actes du Colloque International de Grenoble*, Université des Sciences Sociales de Grenoble, 9-12 May 1989.

[MÉL 92] MÉLÈSE J., *Approches systémiques des organisations, vers l'entreprise à complexité humaine*, Editions d'Organisation, Paris, 1992.

[MOR 02] MOREL C., *Les décisions absurdes : Sociologie des erreurs radicales et persistantes*, Gallimard, Paris, 2002.

[PEN 92] PENNAC D., *Comme un roman*, Gallimard, Paris, 1992.

[PER 99] PERRENOUD P., *Enseigner : agir dans l'urgence, décider dans l'incertitude. Savoirs et compétences dans un métier complexe*, ESF, Paris, 1999.

[PRÉ 98] Préparer l'entrée de la France dans la société de l'information, Programme d'action gouvernemental (PAGSI), La Documentation Française, Paris, 1998.

[ROS 96] ROSNAY (DE) J., *Le Monde Diplomatique*, août 1996.

[SEN 02] SENGE P., *La 5° discipline : L'art et la pratique des organisations intelligentes,* 212-213, First, Paris, 2002.

[THÉ 94] THÉRY G., BONNAFÉ A., GUIEYESSE M., *Les Autoroutes de l'information : Rapport au Premier Ministre*, La Documentation Française, October 1994.

[VAR 93] VARELA F.J., THOMPSON E., ROSH E., *L'inscription corporelle de l'esprit*, Le Seuil, Paris, 1993.

Chapter 2

Management 2.0

2.1. Introduction

We are living in a 2.0 world characterized by an increase in our capability to participate, to contribute to public interest, to share and increase knowledge, to help other human beings to overcome difficulties, and to provoke political, social and cultural changes. In sum, we can assert that we enjoy the freedom to integrate different projects and play distinct roles, depending on nothing but our willingness to participate.

Without ever denying all the freedom that the 21st Century offers us, this chapter will analyze the extent to which present-day organizations see this new social reality. We state in advance that such a reflexion is of great importance not only as a philosophical question, but also, and particularly, for understanding that it is possible to make efficient, effective, and intensive use of new organizational practices which spring from social evolution associated with new technological potentialities. In particular, we will address the competitive intelligence issue, a question that implies that organizations should be in tune with their time by developing models that allow us to take advantage of potentialities of mass production, collective intelligence, and collective knowledge.

Chapter written by Miguel ROMBERT TRIGO, João CASQUEIRA CARDOSO and Bruno Filipe CARVALHO SOARES.

Researchers like Thomas Friedman [FRI 05] asserted that "the world is flat". Henry Mintzbergand Gary Hamel had already addressed, some years ago, the need for companies to innovate in the way they organize themselves and in the way they carry out their daily activities. It is only under this condition that they would be able to survive in this completely new world that we live in. Hamel and Breen [HAM 07] and Skarzynski and Gibson [SKA 08] evoke the imperativity of radically innovating in the way management is applied, by indicating the necessity of making innovation not to be seen as a magical word, but that it should be a mandatory part of all organizations' documentation, so that it can really be part of its DNA by becoming a core competence – that, being practiced by all, should become the real engine behind changes that occur. Hamel and Breen [HAM 07], on their part, are in support that a Management 2.0 should exist in order to cope with the challenges of the World 2.0.

Within this logic, we defend the need for organizations to sustainably adopt a beta model – one of the foundations of the 2.0 concept – and to constantly reinvent themselves in order to take advantage of the opportunities offered by the market. Therein lies the key to successful implementation of an innovation culture that runs throughout the whole organization. Therefore, the potential of efficient application contained in all the fundamental theories and practices for the success of organizations in the 21st Century would be better exploited.

We include in these routines and *modus vivendi* of organizations adapted to the 21st Century, the use of new management practices, application of competitive intelligence, and the use of innovation at all organizational levels (and as one of its core competences). Finally, we include education in companies as catalyst of consolidated changes in training and in preparation of people, and by the adoption of a new strategic business mentality.

2.2. Competitive environment of the 21st Century

To discuss the concept of Management 2.0 and its importance for widening of competitive intelligence (as fundamental strategic ally for companies that want to be successful), we will begin by following the opposite path. We will begin from the concept of Competitive Intelligence 2.0 (that provides an analysis of the competitive environment surrounding all organizations in the 21st Century which indicates the principal technological and social tendencies). It is therefore possible to better justify the need for changes in management practice.

2.2.1. *Competitive environment*

While observing the competitive environment surrounding present-day organizations, we are faced with the following challenges:

– Increasingly fierce competition: it is becoming more and more difficult to identify its origin and the manner in which its *modus operandi* can reach us.

– Constant evolution of the market: accessing the market is becoming increasingly easier for new competitors, who do not need a global structure to compete in international markets. On the other hand, the execution agility of these new organizations, combined with the ease of access to quality information, allows the creation of completely new project models (that correspond more efficiently to the needs of consumers), which may question the leadership of "traditional organizations" or even monopolize a market.

– Increasing precariousness of organizational strategies and competitive advantages: it is easier to have access to information than to successfully copy and adopt winning strategies, which signifies that those who, one way or the other, have indicated a success pathway quickly, lose their advantage.

– New technological and social trends: which we have to adapt to and find a better way of using in order to be armed with the best tools and the best people so as to be more rapid and efficient than the competitor in satisfying clients' and consumers' needs.

2.2.2 *Technological trends and innovations*

2.2.2.1. *Intensive use of Internet in mobile devices*

The major technological innovations that we are experiencing today are partly due to the Internet – its intensive and inclusive use. These innovations are equally influenced by the increasing portability of mobile devices – laptops, tablet PCs, and smart phones – through which it is possible to connect to the Internet, hence allowing us to always be online, to carry out a growing multiplicity of actions. This reality is clearly confirmed by Figures 2.1, 2.2, and 2.3, showing the global evolution of Internet penetration and the number of Internet users as well as the number of mobile phone subscribers.

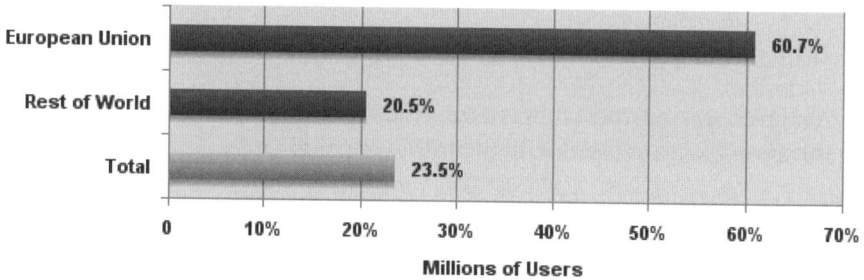

Figure 2.1. *[INT 09] Internet penetration (December 2008)*

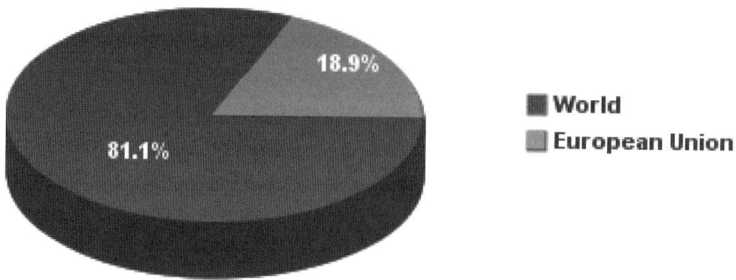

Figure 2.2. *[INT 09] Internet users (December 2008)*

Internet is meant not just for negotiating goods and services, recruiting, searching, learning, having fun, getting information, developing friendship and relationships, but it is also largely responsible for what was formally imaginable for the majority of people on this planet: giving an opinion, having a variety of choices, sharing knowledge, and easily communicating with almost the entire globe. Mobility and participation are the new realities that broaden organizational challenges and opportunities.

When talking about mobility, two realities should be considered: that of organizations and that of (new) consumers. As a member of an organization, we do not need to be present physically at a specific place (e.g. organizational headquarters) any longer in order to have access to all documents and information required for decision-making, to hold meetings with people from different departments of the organization, or to develop a team project in real time.

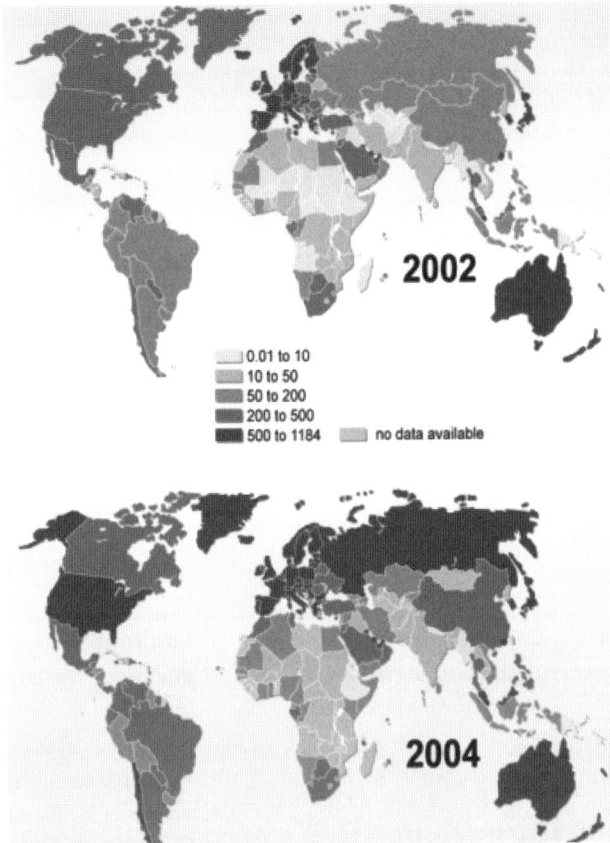

Figure 2.3. *[UNE 06] Number of mobile phone subscribers (in thousands)*

The exponential growth of the Internet, accompanied by the evolution of technical devices, has led to an increase in online services (such as consultation of bank statements, payment via mobile phone, buying of movie tickets) offered to consumers.

This situation provides, on the one hand, great advantages to those who are able to take advantage of its potentialities and facilities – to create new commercial models for example, for improving communication with the internal and external public (clients, consumers and suppliers) of the organization. On the other hand, this reality poses new challenges and opportunities completely different from those witnessed 10 years ago. Some of the identified changes that the Internet and mobile devices currently offer are mentioned in Table 2.1.

Change	Challenge	Opportunities
Internet connectivity through mobile devices	Increase in the possibility of comparing different offers just in time	Possibility of consulting competitive offers just in time in order to better adopt products and services
Technological evolution of devices	Consumer frustration due to unavailability of some services which are already popular (in other regions of the world)	Increase in interaction capability with consumers (e.g. Bluetooth)
Increase in the number of users of devices	Ease of critical information exchange among consumers	Creation of new business opportunities based on user communities which share positive information and reap the dividends of this positive influence
Development of new "buying culture"	Lack of preparation of organizational staff toward new clients and business models	Internet as business model – producing an exponential increase in the number of potential clients/consumers without any required investment in physical structures

Table 2.1. *Principal challenges/opportunities enhanced by the Internet and its applications*

2.2.2.2. *Social trends*

When analyzing social trends, we should start by perceiving the people who currently constitute our society – What are their behaviors? What are the most significant changes to them concerning norms and social values?

This identification allows us to better align the form in which organizations have to structure themselves so that they can be clearly in tune with the people of the 21st Century, and be more sensible and demanding when they have to meet the real needs of a particular society.

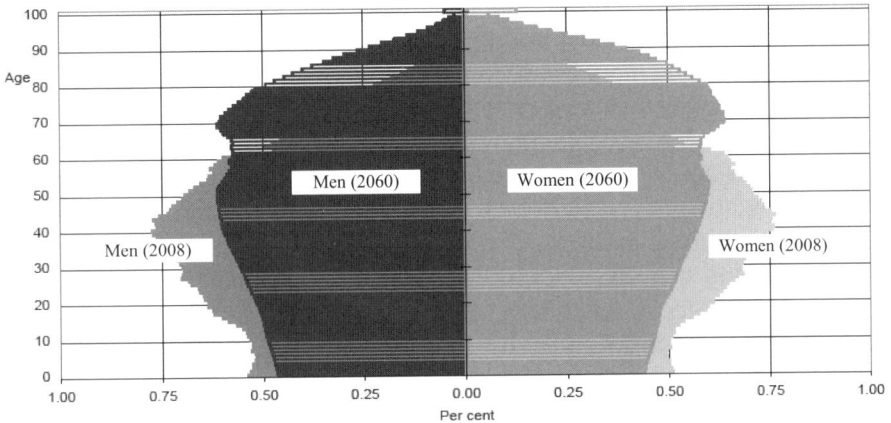

Figure 2.4. *[PIS 09] Age pyramid, UE 27, 2008-2060*

Figure 2.4 shows that the most significant population fringe, presently and in the 27 EU countries, is primarily made up of citizens (men and women) born between January 1946 and December 1997.

It is within this precise time interval that individuals living today and working in different organizations and companies were born. These individuals, following the expression of Tapscott [TAP 08], can be classified as belonging to three different generations (see Table 2.2).

Baby Boom Generation	X Generation	Net Generation
From January 1946 to December 1964	From January 1965 to December 1976	From January 1977 to December 1997

Table 2.2. *Three generations living today and working in markets and organizations (adapted from Tapscott [TAP 08])*

The following developments provide a deeper analysis of the "Net generation". This analysis will be continued by identifying what can be referred to as the "new stronger sex" – i.e. the female population in Europe.

2.2.2.2.1. Net generation

The current generation is the first which grew with constant contact with the Internet. A study[1], at the inception of Tapscott's book [TAP 08], shows that for the first time in the history of humanity, a generation feels at ease, better prepared, and more knowledgeable about this fundamental innovation (Internet) to society than their parents.

Their constant contact with the Internet makes them excellent researchers (seeking information which they require). They develop extraordinary reasoning and investigative capabilities and become extremely critical about the information they have access to. Their mastery of digital technologies allows them to learn, have fun, communicate, socialize, work, and create community groups in a way totally different from previous generations.

Another equally interesting phenomenon is the fact that, through Internet, cultural and behavioral differences which differentiate us based on our geographical situation, training, and experiences tend to become insignificant. For the first time in history, we can speak about a real "global generation" with quite similar habits and behaviors.

But what are these habits and behaviors that characterize the Net generation? Tapscott [TAP 08] calls them the "norms" and defines them as follows: freedom, costumization, scrutiny, integrity, collaboration, entertainment, speed, and innovation.

2.2.2.2.2. The new "stronger sex"

In absolute terms, women are numerically the majority when adressing the 27 countries of the European Union, and they have started becoming leaders in domains that were previously closed against them (e.g. medicine, law, technology, and higher education training). This evolution, according to Peters [PET 08], is due to three fundamental rules that govern our economy: "say goodbye to command and control…, say goodbye to knowing where your place is…, and say goodbye to hierarchy". According to the author,

---

1 "The Net Generation: A Strategic Investigation" is a \$4 million research project, constituting the completion of 1,750 interviews with American and Canadian young people (between 13 and 20 years), and 5,935 interviews with young people (between 6 and 29 years) in 12 countries (USA, Canada, United Kingdom, Germany, France, Spain, Mexico, Brazil, Russia, China, Japan, and India).

organizational necessities "…are becoming more and more compatible with feminine values", which are embodied in the following attributes:

– women have better improvisation capabilities than men;

– women are more auto-determined and more sensitive to truth than men;

– women appreciate and depend on their intuition more than men do;

– unlike men, women envisage natural access to power (non-hierarchical power);

– women understand and build relationships more easily than men.

Penn and Zalesne [PEN 08] speak about changes in behaviors and experiences of women, which have made possible some changes that we see in today's world. Today, a woman can remain single as long as she wishes to. She can choose a demanding profession. She can choose to marry a man younger (with lower professional ambitions) than her. All these are more and more socially accepted. Today, it is therefore easier for women to escape prejudices when they invest in their personal lives. Moreover, the "part-time" couple phenomenon[2] is tending to grow, allowing women to have greater autonomy in managing their lives.

These two authors present other concrete data susceptible to show that women are "actually gaining the ground" in the society and in economy. Therefore, women constitute the majority that use the technology and influence and decide the sales of the majority of domestic appliances which are available for sales in the USA[3].

These new social realities affect organizational life. They should absolutely be considered through the prism of the need for change in organizational paradigms so that the integration of these new actors into the daily life of an organization can be performed in the context of mutual respect and collective construction. On the contrary, there is the risk of fewer positive consequences on the organization.

---

2 Couples who live separately focused on a professional career, and get together only at weekends.

3 Women spend more on technology than men – marginal difference of about 3 to 2. Women are responsible for about 57% of technology purchases. In 2006, they were responsible for nearly $90 billion sales of current consumer electronic products.

2.2.2.3. *Social and technological trends*

Having analyzed the major technological and social trends, it is now time to look carefully into the events resulting from the intersection of the two trends that cause changes in the way institutions should organize themselves and carry out their day-to-day work.

"The most intelligent leaders see clearly that the most important word that needs to be valorized when talking about social networking is the word 'working'. Basically, a new production mode is being born, no more, no less…

In this new world of collaboration, it is normal that couples are formed to create value, often outside the walls of traditional organizations...

With more than a thousand million people interconnected throughout the whole world, collaboration and teamwork have become the engines of the business world" [TAP 08].

The Internet has supported the paradigm shift in communication and business, and has gradually modified our way of thinking in the business world. "Future professionals"[4] will take a dominant place and will bring with them a totally different approach to work. In this sense, institutions wishing to ensure their survival are required to initiate internal processes that allow them to adopt organizational models that can better support the changes occurring around them.

Therefore, we must work toward changing organizational paradigms rooted in institutions for a long time. To succeed, each member of the organization would have to understand how to integrate the best changes; and to understand that if a task has been executed in a way for many years, henceforth, this format will neither be effective nor efficient to achieve organizational objectives and ambition.

---

4 According to the definition of "Brains of the Future" by Daniel Pink [PIN 07]: "…usually creators and empathetic people, model translators and producers of meaning. These people – artists, inventors, designers, storytellers, professionals and volunteers in social, spiritual and health fields, and systemic thinkers – will reap the best fruits offered by the society and share their greatest joys".

Among the activities to be developed, we must invest more time in watching, listening, thinking, learning…

…identifying daily work practices that were previously performed internally which, due to the ease of exchange through the Internet, can be performed more efficiently when they are relocated;

…implementing educational processes in organization, which will provide staff with new techniques and knowledge, and which will allow us to develop new formats for performing daily work e.g. using video conferencing tools for holding meetings, thereby avoiding travel expenditures (in time and money) and allowing exponential increase in the number of meetings conducted in geographically distant buildings;

…using different tools, available through the Internet to the 21st Century workers, by adopting them in the organization;

…working in collaboration;

…adapting to the Net generation in the workplace (from recruiting to motivation: rethink how to recruit, how to reward, how to develop, and how to supervise the talents); the Net generation in the market: know the "Prosumer"[5] (who wants to help you in co-innovating and co-creating products and services).

These are the practices that can (and should) be introduced in our institutions in the future (if we want to). It is essential to consider the changes that we reviewed in this chapter in a positive and active way by not only trying to identify the opportunities that they could offer, but also trying to anticipate and minimize the risks that they could include.

Employees and organizations possessing these abilities and skills will become winners in productivity (in performing tasks) and, they will also be able to intervene in new market niches, which was unimaginable in the 1.0 world.

---

5 Neologism for producer-consumer

2.3. Management 2.0: the world is flat, but organizations should be full

"If things seem to be under control, it means we are not moving fast enough"

Mario Andretti.

The scenarios presented above increase the need for new strategic mentality on the part of organizations, if they have to ensure their sustainability. We believe that the most efficient concept to currently ensure the sustainability of an organization is its anticipation capability: being able to highly anticipate consumers' needs and become the fastest competitor to meet the needs.

We should continually live in maximum search for opportunities so that our know-how and skills can guarantee us competitive advantages. To achieve this objective, we need to build an organization where innovation and competence are integral part of organizational culture. Different steps are to be followed in order to achieve this objective. The experiences of organizations that have already reached this threshold show that it begins with the introduction of innovative processes on how to organize and manage the institution.

The previously identified trends should be apprehended and they must give way to changes in the way we organize ourselves, the way we motivate and "demotivate" the staff, the way we define and introduce strategies and models in the business world and finally, and in the way we perform our daily work which allows us to offer better services and products in order to satisfy the consumers' needs.

For changes to be successful, they must start from the top hierarchy, who then will be responsible for transmitting them to other sectors of the organization. This is the major secret of companies which are born with a different philosophy e.g. W.L. Gore (1958), and more recently Google, Atlassian and Zappos; but also companies that have been able to adapt themselves to their competitive environment e.g. SEMCO, SEI Investments (SEI), Procter and Gamble (P&G), General Electric (GE), Cemex, Lego, Nokia, Cisco and Best Buy. This is the case of organizations that have created innovative management practices, to promote the integration of a new strategic mindset.

We shall now present some of the changes which we think are necessary to introduce if we must follow the steps of organizations which have changed and, as supreme recompense, have successfully ensured a sustained growth of an organization.

2.3.1. *Creativity, sharing of responsibilities, and sharing of results*

To develop a new strategic mindset, we need to think about organizations in a new way and, particularly, to succeed in involving clients and employees at a higher level than usual, by ensuring that these links are materialized through the direct and active participation of these actors in the life of the organization. Intelligent organizations constantly seek the best ways of involving their employees and clients. Thus, they can aspire to create an organization where everyone actually gives his/her best, contradicting the results of the 2007-2008 Towers Perrin Global Workforce Study[6] [TOW 08].

To change this reality, we must implement a series of changes, starting with the form in which most of the organizations are structured (physically and mentally). Organizations have to essentially discharge from the usual departmental boundaries and operate as a team which makes consistent use of the individual strengths of each employee. The goal of this change is to create a highly competitive team (through the diversity of experiences, opinions, skills, knowledge, and information), which will be able to create products and services that meet the needs of its clients and also deep social needs – not only for a short-term benefit.

The ultimate goal of this change[7] is to give the employees enough freedom to apply all their potential to the benefits of the organization. The growth due to the proactive attitude of the employees requests a fair sharing of financial results.

2.3.2. *Define leadership*

Since he became the CEO of P&G in June 2000, A.G. Lafley established that the engine of the organizational growth was based on innovation, which creates added value products and services for consumers. The definition of

---

6 This study carried out with 90,000 employee respondents, indicated that only 21% of them consider themselves being really involved with their organization.

7 By forming a real community in the organization.

customers as P&G's "new CEO" has altered the whole strategy of the organization.

> "Real buyers and users of P&G products constitute a rich source of innovation, if we simply pay attention to listening to them, observing them in everyday living, and even living with them. We must win two moments of daily truth: first, on the shelves at stores when they decide whether to buy P&G product or that of the competitor; the second moment comes up in their houses, when they and their family use our products and decide whether or not we have kept our trademark promises. It is by constantly winning on these two points, everyday, that we succeed in gaining our customers' loyalty and ensure short and medium term growth of the organization" [LAF 09].

According to Lafley and Charan [LAF 09], P&G has invested more than $1,000 million, between 2002 and 2007, in changing a traditional market study department into a center for understanding consumers – generator of knowledge from more than 4 million consumers per year.

> "This total immersion leads to gaining richer knowledge about consumers, which helps in identifying opportunities that are often not understood by traditional investigation" [LAF 09].

On a first look, this report could make us think that we will never have enough money to invest in consumer studies. It is better not to reason this way. We should rather think that if a company like P&G could invest heavily in this domain, we ought to do likewise. In this phase, the difficulty resides in understanding how to follow good examples and adapting them to our reality. That is our suggestion.

Coming back to technological and social trends, previously discussed, it is possible to identify some solutions. Let's look at what happens in some companies, using two examples from Libert and Spector [LIB 09]:

– research carried out in 2007 by Forrester Research with 119 senior managers, responsible for competitive information, presented the most surprising result that "89% used at least one of the six technologies for information gathering from the public, including tools as improbable as podcasts, wikis, blogs, and social networks";

– other research carried out at a global level by McKinsey & Company, with 280 senior managers "discovered that the most advanced country in this domain is India, with 80% of the companies foreseeing an increase in expenditure with online groups in the next 3 years. North America, with an increased forecast of 65% of the companies, came third following the Asian-Pacific companies".

This is the first step toward change! Considering that it is the customer who should drive our actions as an organization, alters a whole series of paradigms which normally prevent us from increasing organizational growth exponentially – like, believing in people, enabling interactions and, sharing of information, knowledge and know-how.

2.3.3. *Collective and participative intelligence: employees' involvement*

When talking about employees' involvement, the most popular model is the process run by Toyota, which derives directly from *Kaizen*[8]. *Kaizen*'s main refrain is the creation, from individuals who do the work, of patterns that must be monitored and continually reinvented to achieve actual improvements in daily work. A good example of this capability to use each employee's knowledge is shown in the report that during 2005, Toyota used more than 540,000 improvement ideas obtained from its employees.

As we progress in the 21st Century, employees' involvement models are becoming increasingly effective and efficient. Here are two examples:

– Best Buy – Costumer Centric Cycle. This is a bottom-up systematic innovation program, which invites everyone to dream up new ways of creating differentiated shopping experiences from the beginning to the end. The gains from this initiative are shared, through financial rewards paid to participants and, through the "real time" changes that Best Buy had implemented in all its sectors. But, most importantly these changes are driven by motivated employees throughout the organization and not by the Board of Directors;

– IBM. In agreement with Skarzynski and Gibson [SKA 08], the company started an effort, focused on recent years, to involve all its people in a global discussion on organizational strategies and objectives. In an

---

8 *Kaizen* is a Japanese word, meaning gradual and continuous improvement. It entails presentation of ideas, and not necessarily their acceptance. A real business incubator, *Kaizen* is a practice that promotes a strong sense of laboratorial curiosity in companies like Toyota. It is a proven basic way of collecting human creativity [MAY 06].

online session of 72 hours, all 319,000 IBM employees worldwide were invited into a kind of value improvisation through the company's general Intranet, to exchange views on the factors that could differentiate the company in the market and guide the actions of every employee.

Despite the quality and depth of the measures stated above, we wish to highlight the examples of W.L. Gore and SEI Investments in the USA, and SEMCO in Brazil, for their completely innovative way of developing a culture of deep involvement of employees.

2.3.3.1. *W.L. Gore*

The company W.L. Gore, established in 1958, was born with the "zero bureaucracy" philosophy, which was influential in developing a completely original organizational and management model that breaks with traditional paradigms. "The singularity of Gore is the fact that it is as innovative in its operational principles as it is in various product lines" [DEU 04].

The company has been regularly voted as one of the best organizations to work with, thanks to the "Gore method" which is based on the following principles:

– horizontal organizational structure. It is difficult to find such a thing as hierarchy;

– everyone belongs to the same category and has the same professional title "associate";

– there is neither a chain of commands nor predetermined communication channels (everyone can speak to anyone);

– teams are organized as small task forces;

– team size is limited (150 to 200 people) so that everyone may know each other, know what others are doing, and know who is competent and skilled enough to perform a determined task;

– there is no boss. There are only leaders;

– your team is your leader. You should not disappoint your team;

– once employed at Gore, new employees can rely on a mentor, as they are not assigned to the leaders. They have to search for ideas which can be interesting to the company and for him or her to do. Associates choose the leaders who they want to follow;

– associates have to spend 10% of their time looking for new ideas;

– performance evaluation is based on a peer assessment system;

– there is a principle of distribution of benefits to employees, or other types of stimulation, directly related to the company's results.

Some benefits already verified from this model include: the steady growth of W.L. Gore product line (which can be found in sectors such as industry, advanced technology and electronics, medical products, etc.), and the expansion of an enviable network of clients.

2.3.3.2. *SEI investments*

SEI Investments, another organization that is distinguishable in the way it succeeds in involving its employees, presents other interesting concepts. In SEI, any type of job must be done as a team.

> "…we do all our work as a team, it is more important than having people to command. When people fail to progress in the company, it is not because of their boss but because of their colleagues" (Al West, CEO, SEI Investments).

To implement this philosophy, some important measures were taken by SEI, as explained by [KIR 07]: "all furniture is on wheels so that each employee can create his/her own workspace". Added to this is the fact that there are no partitions (there are no individual workspaces) and nobody has the right to have a personal administrative assistant. All these create an extraordinary pyramidal informal atmosphere.

Even more interesting is the fact that SEI has absolutely no organizational chain of command. The business unit par excellence is the team. Some teams are permanent (in proportion to the importance of the customer or the strategic importance of the market), but a lot of teams are temporary with a lifespan exclusively limited to the time required to provide solutions or resolve a problem. As a result, SEI is an ever-moving organization that succeeds in adapting to the needs of the moment.

"We call this the management fluid. People understand the domains they are good in, and this discovery defines the role they are going to play". [WES 07].

2.3.3.3. *SEMCO*

In discussing SEMCO, we refer to the "revolution" that began to take shape in 1986, at the initiative of the Brazilian entrepreneur Ricardo Semler[9], when he decided to put an end to a centralized management system and started developing all the initiatives of SEMCO based on the concept of business units. These changes, gave greater freedom and huge responsibility to all the managers and employees in general, and were the result of the innovative manner in which Semler thought and theorized management.

To every dogma presented in the theories of management, Semler presented another vision based on the knowledge from other sciences which have performed a phenomenal job on the study of human behavior.

Semler [SEM 88] identified the four principal factors that influence the sustainability of an organization:

– timely perception of the need for changes and the courage to implement them before it is too late;

– running the company with effective participation of the employees;

– being flexible in administration and open to transformations;

– possessing a well-defined culture and conforming to its fundamental principles.

Unlike other organizations, SEMCO promotes job security as a necessary condition for motivation, productivity, and long-term survival. This differentiation was of the utmost importance so that the participatory management at SEMCO (the effective participation of employees in the life of the company) could be a reality.

"Disassemble the corporate apparatus with centralized systems and other dinosaurs. Release the clamps and let the staff work freely and independently in small units. Lose a few

---

9 Also the author of the world best-seller *"Maverick!: the success story behind the world's most unusual workplace"*.

hours of sleep at the beginning, to earn the equivalent of two Valiums per night in a year" [SEM 02].

This new way of involving people in organizational life is materialized with the implementation of measures such as those presented in Table 2.3.

Programs for employees participation in decision process	Employees' participation in management meeting. Groups monthly meetings to create the list of necessary improvements. Greater involvement of work councils. Setting up committees to improve working environment.
Increase rate of enterprise/employees credibility	Company's financial information and workers salary information available via Internet. Participative evaluation and recruiting processes.
Working environment basic conditions 101	Defined by those doing the work. Disclosure of all the costs involved through the Internet.
Weak turn-over	To ensure the continuity of the participation culture.
Managers and directors compromise	Insofar they are the ones that would most feel a loss of status, they should be the first ensuring the participation of all.
Employees training	A well-educated workforce is essential for more active participation in the life of the organization.
Profit sharing	Absence of paternalism. No assignment of benefits without profits. Democratic distribution of possible prizes.
Freedom to choose time schedule	The recruited individuals must meet targets. If they simply meet their time schedules, the company will be nothing more than an internal college.

Table 2.3. *SEMCO participatory management*

By creating mechanisms that involve employees in the institution's life[10], organizations will increase their capability to successfully meet the expectations of customers.

If we think in an objective way, we would realize that those who best know an individual (apart from himself/herself) are those living with

_____

10 In order to promote and increase taking into consideration of their ideas and knowledge in operational strategies and models.

him/her. Transposing this scenario into an organization, we realize that those who best know the customers (apart from themselves) are the employees who are in contact with them every day.

2.4. Conclusion

We have illustrated in this chapter that organizational management cannot remain unchanged, given the everyday realities of the world, and that employees should be allowed to really become an integral part of the organization. We have also described with some examples, what some companies (unfortunately fewer than would be desirable) are already doing, with good results. The next step is up to you – go ahead, fasten your seat belt, and get into 21st Century velocity.

2.5. Bibliography

[DEU 04] DEUTSCHMAN A., "The fabric of creativity", *Fast Company*, vol. 89, p. 54-60, 2004.

[FRI 05] FRIEDMAN T., *The World is Flat: A Brief History of the Twenty-First Century*, 3rd edition, Farrar, Straus and Giroux, 2005.

[HAM 07] HAMEL G., BREEN B., *The Future of management*, Harvard Business School Press, Boston, 2007.

[INT 09] INTERNET WORLD STATS, http://www.Internetworldstats.com/stats4.htm#graphics (accessed 30 June 2009).

[KIR 07] KIRSNER S., "Total teamwork - SEI investments", *Fast Company*, December 2007.

[LAF 09] LAFLEY A., CHARAN R., *Controle as Regras do Jogo*, Actual Editora, Lisbonne, 2009.

[LIB 09] LIBERT B., SPECTOR J., *Muitas Cabeças Pensam Melhor*, Lua de Papel, Alfragide, 2009.

[PEN 08] PENN M., ZALESNE K., *Microtrends*, Lua de papel, São Paulo, 2008.

[PET 08] PETERS T.J., WATERMAN R.H., *In Search of Excellence*, Harper Collins Publishers, London, 2008.

[PIN 07] PINK D., *O cérebro do futuro: a revolução do ladodireito do cérebro*, Elsevier, Rio de Janeiro, 2007.

[PIS 09] PISON G., *Population & Societies*, Ined, no. 454, March 2009.

[SEM 88] SEMLER R., *Virando a própria mesa*, Best Seller, São Paulo, 1988.

[SKA 08] SKARZYNSKI P., GIBSON R., *Innovation to the Core*, Harvard Business School Publishing, Boston, 2008.

[TAP 08] TAPSCOTT D., *Grown Up Digital*, McGraw-Hill, New York, 2008.

[TOW 08] The Towers Perrin 2007-2008, Global Workforce Study: Insights to Drive Growth, HCI White Paper, 30 October 2008.

[UNE 06] UNEP/GRID-Arendal, Mobile phone subscribers, UNEP/GRID-Arendal Maps and Graphics Library, 2006. http://maps.grida.no/go/graphic/mobile_phone_subscribers (accessed 30 June 2009).

[WES 07] WEST A., "Total Teamwork - SEI Investments", *Fast Company*, 18 December 2007.

Chapter 3

Sustainable Development 2.0: Seeking "The Creation of Shared Values"

3.1. Introduction

The present title has been inspired by the "communicative posture" chosen by Nestlé[1] for representing their social responsibility policy – the concept of the creation of shared values – and the ontological concern of sustainable development actors to revisit, at the beginning of the 21st Century, the values of liberal economy and capitalism.

Until the mid-1990s, sustainable development was a concept relatively strange to companies and non-governmental actors. Apart from the good practices of companies engaged very early usually by conviction, the strategy of the executive was the pride of the shareholder model dear to the neoliberals. The shared idea of the managers' responsibility toward their

---

Chapter written by Fabrice MAULEON.
1 The concept of shared value was raised for the first time in Nestle's 2005 report on social responsibility as applied in Latin America. Its authors, FSG Social Impact Advisors, led by Professor Mark Kramer of John F. Kennedy School of Government, Harvard University, found that Nestlé was "in the forefront of the companies that created real shared value for themselves and for the society at every stage of their business processes". Since January 2011, Michael Porter and Kramer, leaders in Strategy, have urged leaders to recognize that "shared value is not social responsibility, philanthropy, or even sustainability, but a new way to achieve economic success". They advocate in an Harvard Business Review article ("The Big Idea: Creating Shared Value" (January/February 2011)) that creating shared value will drive the next wave of innovation and growth in the global economy.

shareholders was to provide them with the maximum possible dividends. Value creation was therefore principally related to the shareholders. Over a decade later, the attention of economic and political journalists, present at international conferences on this theme, focused on the testimonies of traditional or online firms, which revealed some of the practices that are sometimes more successful than their original practice. This enthusiasm has given way to doubt and questions: what's behind the facade erected by corporate communications on sustainable development (window dressing)? How can we understand the enthusiasm for sustainable development and corporate social responsibility? Does the proliferation of Internet sites related to sustainable development result in a new form of exchange and sharing? The synchronism of these questions with the advent of Web 2.0 is not entirely due to chance. It reflects a re-configuration of the world and questionings which are both modern and classic.

These interrogations go beyond the framework of organizations, information and communication technology. They reveal the emergence of a new model for creating shared values and a revisit of matters closely related to the nature of organizations and that of the Web in general.

The June 2007 white paper on *Web2.0 for watch and information retrieval* highlights that "several authors agree on the idea that Web 2.0 is neither a technological nor a social break. It is rather a return to the source, a rebirth of the Web as it was originally conceived, with the users appropriating the creation and dissemination mode. According to Paul Graham[2], the pioneer of the Internet, "Web 2.0 is the Web as it is conceived to be used. The trends that we distinguish are simply the inherent nature of the Web emerging from bad practices that were imposed during the Internet bubble". The idea of revisiting an idea or concept in order to remove bad practices and capitalize on the "good ones", is therefore a shared value between the concept of sustainable development and Web 2.0.

Our study on the emergence of Sustainable Development 2.0, applied to organizations, thus refers to a naturalistic inquiry of the firm through the study of its use of ICT. One wonders on the evolution of the firm: does it become more responsible or more able to better use the progress tools to communicate? But the object of our study may be completed at this level of our development.

---

2 http://www.paulgraham.com/bio.htm; http://fr.wikipedia.org/wiki/Paul_Graham.

It is possible to go beyond this naturalistic inquiry in favor of a simple flat examination of the business communication ethics. This posture becomes more functionalist as it implies going from a pessimistic report while maintaining the view that organizations have no nature. One can also consider it to imply that an organization is determined only by a set of objective factors, as discussed by a number of authors [BOA 02, CAP 04, DUB 02, ELK 97, FER 01, IGA 02, and LAV 02].

In all cases, the idea of bringing out the profile of Sustainable Development 2.0 allows us to advocate for the recognition of a new status for organizations by using Web 2.0: a new mode of operation of organizations which redesigns a social architecture and a shared, collaborative and distributed computer application architecture; geared toward more responsibility and the limitation of its negative social and environmental impacts. Thus, we will initially address common features of a new paradigm of organizational management that is 2.0 compliant before considering the contours of Sustainable Development 2.0.

3.2. Common features of a new paradigm of 2.0 compliant organizational management

Scientists of yesterday and today's ICT actors share the need to attempt to redesign and evolve the organization of our society with enthusiastic advocates of sustainable development – researchers, politicians, businesses, NGOs, etc. Today, new technologies, especially Information and Communication Technology (ICT), appear to be leading the way to more sustainable societal models. In this section, we will not dwell on the characteristics of Web 2.0, which we believe will be broadly covered by other authors of this book. We will mainly focus on the characteristics of sustainable development and corporate social responsibility.

The late 20th Century had witnessed a profound economic, social and political change, which encouraged the questioning of the social role of businesses and highlighted the need for greater openness of organizations to the outside world [BRO 02, GEN 03a, OCD 01, PER 03]. Organizations are required to assume responsibility for the consequences of their activities, to meet expectations or obligations that are not included in the formal law, and to contribute to sustainable development [FRI 02, PER 02].

Research conducted in France on corporate social responsibility oscillates between two dominant positions:

– for some, social responsibility must be related to the ethics and the managers' awareness of the need to integrate extra-economic data in their management decisions (Pesqueux, Igalens). Corporate social responsibility is treated as one of the subtle variations of business ethics of which it would be a collective expression;

– for others, social responsibility is an offshoot of corporate governance. It implies reconciling the views of different stakeholders by adopting a sort of middle path. Responsible investment must ensure a level of performance at least equal in the long run to non-responsible investment [CAS 04, CHA 98, SPI 99].

We can conclude that these two views are in no way conflicting with the character of Web 2.0. Instead they reflect a single reality: the desire of some of the players, on the one hand, to integrate and innovate the broader dimensions in the traditional business model, and, on the other hand, to try to make a participatory approach be the new cement of the social relationship, thus reconciling the main interests of key stakeholders.

3.2.1. *The characters of sustainable development applied to companies*

If we want to distance ourselves a bit from the political definition given by Brundtland and focus on the operational implementation of the concept, it might be interesting to concentrate on the characters and values promoted by the concept.

Villeneuve, in 1998, attempted to synthesize sustainable development in terms of a four-value fundamental:

– socially equitable;

– ecologically viable;

– economically efficient;

– capable of restoring the balance in the North/South relationship and reduce the rich/poor disparity.

Based on these values, we can suggest the following developments. sustainable development (SD) evokes a moral commitment which is based on a balance, which is beneficial to the organization and its stakeholders from the moment it reflects the following combined choices [MAU 05]:

– *An imperatively economically strong enterprise*: this aspect is often overlooked. An organization that integrates this concept (SD) at the center of its activities and its strategy, cannot do so without being economically sound. Bad cases are sometimes quickly made against enterprises that have a strong commitment to SD but are constrained, in order to remain competitive, to arbitrate in favor of the economic aspect and thus lay off employees. But we must not forget that we are concerned here about raising the question on how to promote a sound development and management feature of the organizational activities. The integration of the social and environmental considerations is dependent on the good health of the enterprise. However, this requirement sometimes makes the leaders take difficult steps;

– *A socially integrated enterprise*: this entails that the company ask itself, first of all, how will it integrate the momentum of its employees in its priorities in order to persuade them, out of a common interest, to engage in a medium- and long-term development. Since the English term "social" implies knowing how the company manages the societal impact of its activities in its town, region, country, etc. the firm must measure its societal impact on its intermediate environment i.e. its adjoining municipality, department, region, etc.;

– *An ecologically friendly enterprise*: the fiscal, regulatory or judicial constraints combine to make companies understand that the government and the stakeholders who are sensitive to ecological issues will no longer tolerate what the community has so long endured;

– *A transparent enterprise vis-à-vis its shareholders*: investors and shareholders now combine their efforts to encourage companies to favor a management style that is able to provide information. This commitment to transparency is indispensable for a listed firm and the formalization of the general nature of corporate governance is not limited to the big companies (group of companies) but extends to all organizations.

Social responsibility, as an expression of sustainable development applied to organizations is, therefore, a development process which implies, for an enterprise, undertaking a number of changes which, if they seem obvious, should imperatively be combined to give full effect to this new mode of management.

These elements are not different from the processes observed from Web 2.0 actors since this implies that a firm should:

– *Open its temporal horizon over a long term*. Without exclusively aiming at protecting the interests of future generations, it is demanded that organizations adopt a strategy that is less "short-termistic" and more long-term or at least medium-term oriented. The concept of risk illustrates this need for change. We live today in a "risk society" [BEC 01] which no longer resembles what the actors of the industrial revolution of the 19th Century could have imagined. Since the Bhopal, Chernobyl and other disasters, it has been widely accepted that enterprises are faced with new risks (chemical, industrial, alimentation, etc.) which demand that they should apprehend, as early as possible, all proactive processes likely to fight these new constraints;

– *Take into account its real spatial horizon*. The phrase "think globally to act locally" is now well known. It shows how important it is to no longer consider an enterprise as a single entity but as a local economic structure which cannot escape its local, regional, national and international environments. The strong mobilization of enterprises at Johannesburg perfectly illustrates this changing spatial horizon. The combined action of the World Business Council for Sustainable Development (WBCSD) and the International Chamber of Commerce (ICC) toward the creation of the Business Action for Sustainable Development[3] (BASD) explains how sensitive enterprises are to the sharing of all possible experiences of SD strategy implementation, at an international level. Sustainable management is a concept with an international dimension;

– *Integrate multi-partnership and interdisciplinary approaches*. The notion of stakeholders is a significant example of this character. It occupies a central position today, as a management concept. It entails the activity of "all groups or individuals which can affect (or be affected) by the realization of the objectives of an organization[4]". It is the set of people interested in the good health of the organization. It entails "evaluating the economic, social and ethical effects of a decision on the society before taking such a decision" [MER 99]. An executive, who finds out that his organization is being led by

---

3 A Website, www.basd-action.net, relates these initiatives; especially when they include partnerships with public institutions, local communities, NGOs, etc.

4 We owe this concept to R.E. Freeman, Professor of Business Administration, turned philosopher in the book "Strategic Management: a stakeholder approach" [FRE 84].

interests other than those of the management team and shareholders, must necessarily adopt a partnership and interdisciplinary or transdisciplinary approach.

By considering the distinctive qualities of CSR and SD, through the idea of "sustainable management" the managers would be able to launch a global reflection in directions that are very rarely used in traditional management problems. This offers a better knowledge of the demands of the society with regard to the organization. It emphasizes the need to consider the potential risks of conflicts that an organization symbolizes. This enables the managers to account for "sustainable management", which is a very dynamic strategic reflection due to its multiple stakes. This type of management is therefore global, international and transversal. It could also be given the 2.0 qualitative. Indeed, the new actors of SD (non-commercial enterprises or organizations) concerned with this new mode of management would react as they compose a new civil society of opinion.

3.2.2. *The emergence of a civil society of opinion*

The notion of civil society is not new. It dates back to Antiquity and Aristotle developed it to describe a public space in opposition to *oikos* who described the private sphere. The distinction proposed by the Greek philosopher was based on the relationship with authority: a private sphere corresponds to a vertical power. This enlightens us even today since the civil society of today is dominated by a horizontal and contractual relationship i.e. without a vertical authority [LAI 05].

Today, the modern concept of civil society is traditionally related to the political theories of the state and the former cannot exist without the latter. It is therefore not "normally" possible to think of the civil society independently of the state. Nevertheless, the civil society is becoming more and more constructive as compared with the enterprises, and without the state.

Hence, if we combine the definition of the civil society given by Habermas [HAB 77], who describes it as an autonomous sphere of decision made without the state and market influencing, inflecting or thwarting national or global collective choices, with the finding that civil societies are in conflict with major international firms, it will be necessary to examine later in our development the place and role played by Web 2.0 actors when they take a stand against organizations in the society. Developments relative to the emergence of Activism 2.0 will contribute to this.

In the meantime, we can make the observation that the most sophisticated civil society can develop only in the most sophisticated political societies: those dominated by new technologies, commercial enterprises and financial markets. All actors of the civil society have now become actors of opinion.

The emergence of these new power relationships between enterprises and those affected by their activities is now considered by international firms as a new risk factor. We shall see that the emergence of a militancy (activism) using Web 2.0 can be worrisome in many respects to the economic actors. At the forefront of these dangers is the reputation and image of the company which demands today to be managed as a brand, and especially as a risk. The civil society relayed by the media has come to accept that all these companies penetrate into the intimacy of families, homes and businesses themselves. For this intrusion, the price to be paid is high, since the society expresses itself and claims a certain opinion. This typically manifests itself in a reactive and protesting way and in favor of more accountability and transparency [DHU 05].

Various reasons facilitate the erection of this rampart of opinion by the civil society against the irresponsible companies.

Firstly, and this is the issue that concerns us, the final years of the 20th Century introduced many technological changes. Since the first industrial revolution, social scientists have focused their attention on the central role played by technological advancement responsible for economic growth. We will illustrate this thesis in the second half of our explanation. This disrupts the organization of organizations and influences the relationships between individuals, groups and organizations in which they work [TUS 90]. Henderson and Clark [HEN 90] have shown that technological changes affect the population, industries, organizations and environmental conditions. This is already causing significant changes within organizations and is giving rise to new forms of organizational practices in order to exploit these new developments [POW 96].

This question of interaction between NICT and organizations has resulted in many academic works in the 1980s: "Technology moreover has been at the heart of the theory of organizations beginning with Taylor (1911), Bernard (1938), then March and Simon (1958), Perrow (1967) Thompson (1967), Child (1972), and more recently Barley (1986)" [TUS 90].

Secondly, the place occupied by "enterprises" in the society has also contributed to their media and social exposure. But they are not the only one projected to the front-stage: their detractors have accompanied them as well. On the issue about the importance of taking into consideration the "enterprises", a shift could be noted between the respective perceptions of economists and civil society. In economic theory researches, enterprises have only occupied a congruous place until very recently. "Historically, economic science always had difficulties in understanding organizations and had to gradually abandon the postulates of classical economics, in order to do this. Indeed, the vision of the firm, for example in the theory of general equilibrium in economics, remains insignificant: it is assimilated to an individual agent, without consideration of its internal organization or its own resources. For a long time, economic science has regarded enterprise as a black box and has only considered its behavior as a unique model of profit maximization, i.e. the best use of technical and human capital to get the maximum benefit" [PLA 03].

However, actors of civil society have not suffered a similar myopia. In order to focus on the organization either by working for it or by working with it, stakeholders of the enterprise, whether employee, shareholder, supplier or visitor, have developed a conscience that they have exercised over the firm. Its legitimacy is at the heart of this examination of opinion and this control continues to grow. "In the next society, the greatest challenge to be faced by major enterprises – especially the multinationals – will probably be their social legitimacy: their values, missions, and visions" [DRU 01].

Thirdly, the civil society of opinion is also the result of a crisis of proper conscience at the end of the 20th Century. Web 2.0 is an initiative of the 21st Century, but it also places a bridge between the past century which witnessed the birth of the Internet and the current decade, which houses its use. It is as such, a reaction toward the end of the century.

The 1990s, described by Nobel laureate in Economics Joseph Stiglitz as the Roaring 90's with reference to the Roaring 20's – crazy years that preceded the crash of 1929 – were characterized by an exuberant stock market which was accompanied by excess figures:

– an explosion of companies' executive compensation especially of their variable portion backed by stock options;

– the emergence of new financial instruments and new accounting techniques which allow for skilful manipulation of the accounting or transparency, relative to the actual extent of debt to the balance sheet of the company. These techniques provided a strong temptation to choose to improve the results of the company;

– a deregulation, particularly in the banking sector, which weakens the institutional mechanisms of control when the rules are eased.

Cases like Enron (2001), Andersen (2002) and Parmalat or WordCom (2003) have wronged many shareholders, retirees, creditors, suppliers and employees.

BUSINESS WEEK *and the denunciation of the expectations of stakeholders by the media*

For many years, the media in general and the press in particular, have been an effective remedy for citizens against abuses of power. Indeed, the three traditional powers – the legislative, executive and judicial – may fail, make mistakes and commit some errors. These are, of course, much more pronounced in authoritarian or dictatorial state. But these imperfect nations have no monopoly on the abuse of power. Democratic countries can also commit serious abuses and their journalists have often regarded it as a major duty to denounce these violations. This is why the media is considered as the "fourth power". This opposition force is actually working in favor of the civil society to criticize and democratically oppose illegal decisions which may be unfair or unjust.

In the last few years, the American magazine "Business Week" has become famous by regularly denouncing the iniquities of the capitalist system. This denunciation sounds like a loss of confidence in the concept of enterprise.

Thus, in the special issue, which became famous, the Business Week of September 11, 2000 titled "Too much corporate power" (the boss has too much power), had published a survey that had revealed citizens' views on U.S. firms:

– for 72% of the citizens, companies have too much power on various aspects of people's life;

– for 75%, enterprise executives earn too much;

– for 74%, the big companies have too much influence on the politicians.

This revelation was illustrated by a comment from Alan Greenspan, the director of the U.S. Federal Reserve: "the inability of the capitalist economy to better redistribute wealth may cause a resurgence of protectionism, regulations and intervention of the State".

Business Week has thus continued its denunciation activity throughout the weeks following this first article. The subjects were divers but all the articles were focussed on the excesses of the enterprises:

– November 6, 2000: report denouncing income per capita between the North and South America (Global Capitalism);

– September 18, 2000: report to find out who really benefits from tax cuts (How compassionate is Bush's tax plan?);

– September 18 and November 13, 2000: report on the financing of political parties by private companies in the 2000 U.S. presidential election campaign;

– October 2, 2000: report on Wal Mart and its sweat shops in China;

– November 27, 2000: report in a special issue on workers who were sold as slaves and child labour (Workers in Bondage);

– throughout 2000: several reports on the petrol cartel.

All these factors have led to abuses, scandals and a trust crisis detrimental to the proper functioning of stock markets. They have mostly given rise to a wave of distrust of public opinion which obliges companies to manage their image as they had learnt to do for their brand. In the eyes of company executives, the image and reputation of the company have become predominant elements. They are found throughout the life cycle of the business: downstream, at the recruitment of new employees, throughout the activity during the signing of new partnerships, when considering the turnover and the stock price; upstream, at a price decline due to a boycott organized by an NGO.

Thus, issues relating to sustainable development and corporate social responsibility have placed the notion of reputation and its corollary of reputational risk, at the forefront. One no longer messes with the image of the company since the Nike, Enron or Buffalo Grill cases. Thus, as we shall see more precisely in the second half of our development, that activism can also assume a 2.0 format. If it spreads and is becoming widespread, it should be included in the reputation management, otherwise it is likely to cause major problems for companies in the future. Activism 2.0 is already adopting the following characteristics: diffuse, impulsive, very responsive to current events and evolving – as much synonyms of adjectives of complexity for those who want to stop it. It will be difficult for governments and enterprises to predict, anticipate a Web that is a bottomless pit and react in time to these activist initiatives. We see that power redistribution and social regulation are underway.

All these elements are found in the outlines of Sustainable Development 2.0.

3.3. The outlines of Sustainable Development 2.0

It is often hard to distinguish between these two concepts, namely sustainable development and corporate social responsibility, because they are often confused and alternatively cited to refer to each other [MAU 07]. SD may nevertheless be distinguished from CSR, which represents "the voluntary integration of social concerns and companies environment in their business operations, and their relationships with all internal and external stakeholders (shareholders, staff, customers, suppliers and partners, local authorities, associations, etc.) in order to fully meet the applicable legal obligations, investing in human capital and respecting their environment (ecology and territory)[5]". But we will not distinguish between these two concepts in our development. Thus, if the concept of CSR is simply the implementation of the concept of SD by companies, the idea of Sustainable Development 2.0 could be revealed by putting into perspective the Web 2.0 and various impacts of the companies' activity.

---

5 Source: Green paper of European Union Commission–July 18, 2001.

3.3.1. Web 2.0 helping the social impact of an enterprise

Web 2.0 is considered as a revival of ICT. This evolution seems to concern both the technology itself and the use to which it is put, which promotes a community approach for the advantage of stakeholders, as well as the activists.

3.3.1.1. *The determinants of ICT for stakeholders*

Since the end of the 20th Century, moral and societal pressures which weigh on large companies are increasing. This phenomenon happens to be the expression of new expectations of the civil society. This movement has been analyzed by researchers. Many academic developments have emerged to describe this alternative perspective to the financial approach of companies. These researches raised the theoretical field of this moral pressure on the firm [MAU 07]. Some of the authors have worked to make readers understand the relationship between market, business and civil society. Thus among the concepts selected, the stakeholder theory can be considered as an essential development to better understand the thematic of corporate social responsibility or sustainable development, both of which raise the question of extra-financial disclosure of large enterprises. The purpose of our present development is not to carry out an academic literature review on this theory. However we will recollect in a few lines, the writers who have made this theory popular. E.R. Freeman [FRE 84] defined stakeholders as "any group or individual who can affect or be affected by the achievement of the objectives of the company". The ambiguousness of this definition [BRO 04] has however led to the establishment of several topologies for the purpose of clarification.

Carroll [CAR 89] made a distinction, now classic, which enables us to identify two major categories of stakeholders:

– the "primary" or "contractual": this refers to actors who are in a direct and contractually determined relationship, as its name suggests, with the company;

– the "secondary" or "diffuse": this refers to actors located around the company on whom the company's action is impacting without being in any contractual link with the company.

Mercier [MER 99] in his own point of view proposes considering that stakeholders are "all agents for whom the development and health of the company are important issues".

In our study, we will maintain that stakeholders formulate demands to organizations in the form of questions which can also be considered as pressures. They can, of course, enjoy taking advantage of the Internet to further and precisely spread their message. The impact of ICT has become crucial to these stakeholder activists.

Moreover, through the development of new technologies, stakeholders are able to interview many experts on issues concerning a company. At any moment, they are aware of a new accident or incident which may affect the company. They can also exchange, compare and thus strengthen the foundations of their actions.

Thus, for example, shareholders can now question in real time leaders who are heads of companies in which they hold shares. They can summon them to answer questions without having to move. This innovation also reinforces the power of control. Similarly, institutional investors also benefit from advances in telecommunications, both in terms of their knowledge of the enterprise and its environment. Consumers may also find themselves on a Website to estimate the feasibility of a group action (www.ensemblenjustice.org or www.classaction.fr). This control can also take the shape of a new form of activism.

3.3.1.2. *Sustainable Development and Activism 2.0*

A new form of Activism 2.0 in the form of an advocacy efforts program for sustainable development, which uses Web 2.0 community networks to prove their point of view or disagreement, is emerging [MAU 10].

Unlike their traditional counterparts, these activists do not have an *a priori* relationship with any militant association or NGO. They are simply User Generated Content (UGC) which promotes consumer activism, often in a flash mob on a given topic. This flexibility is also an apolitical stance. Far from the famous phrase of Stephen Decatur "My country, right or wrong", this form of activism also revisits the traditional and unconditional affiliation to the classical representative bodies: unions or institutional policies. While the latter work on the basis of a fixed membership coupled with a membership fee and calls for a voting discipline from their member, the new activists of

Web 2.0 are most often free, anonymous and their commitment is timely and spontaneous. Everyone can join virtually or temporarily and activism is placed before activist. This activist community is therefore constantly changing and rebuilding its form.

This form of "flash mob[6]" – gathering a group of people in a public place to carry out actions agreed upon in advance before they disperse quickly – seems to be a feature of this new 2.0 form of activism. Indeed, the place, time and shape of the gathering are generally disseminated through the Web and via social media in particular. Digital innovation is therefore crucial. Although in most of the cases participants do not know one another beforehand, these gatherings federate efficient, fast and stealthy clusters, since all members of the group are dispersed quickly, once the message is broadcast and before the arrival of public authorities. In France, these kinds of rally were held recently to challenge, for example, the DADVSI bill.

The means employed are those of Web 2.0. Thus, the main collaborative community exchange networks – MySpace, Facebook, Twitter, Digg, YouTube and various blogs–allow usage of the Internet to build an activist approach. This is not necessarily synonymous with contestation. It can also take the form of an initiative to promote the good actors of sustainable development, while distinguishing them from their classical competitors (see the section below on Carrotmob).

Carrotmob or choosing a Regulation 2.0 by reward

Carrotmob is an activism program in support of sustainable development using Web 2.0 community networks to create competition between traders in a neighborhood, based on an environmental progress objective, and then create flash mob shopping purchases to reward the most deserving merchant.

At present, the initial operations mounted by Carrotmob are strictly microlocal – at the level of a quarter. They target local shops which were direct competitors. But one can easily imagine an extension.

First, Carrotmob chooses small local businesses. It offers them a deal: make the best possible offer of sustainable development progress in their

6 http://fr.wikipedia.org/wiki/Flash_mob.

daily activities: it may be in the form of an auction based on the percentage of sales that the merchant is willing to reinvest, for example, in energy saving work (first founding experience of Carrotmob at San Francisco); or filmed statements of intentions (second recent initiative).

Secondly, the offers of traders are published on the Carrotmob site. Internet users (local and distant) may be called to vote for the best credible offer.

Thirdly, the company elected as the most virtuous will benefit from a magic day: the day when all consumers involved in this process come to buy in the store to make it rain (money rain). The festival, the accompanying musical concert, the pleasure of being together in large numbers – all these are part of the reward for the participants who do not earn any material gain (they even pay for consuming based on order) but a good social conscience.

Finally, Carrotmob provides the recipient with a board of engineers in sustainable development to carry out the improvement works to which he is committed.

Source: www.carrotmob.org

The maxim "Think global, act local", dear to the partisans of sustainable development is central to these initiatives. This feature explains itself in the North American cultural background of these sites. Indeed, English-Speaking activism often takes the form of "small-step" activism. It differs, in general, from the continental activism in that their actions are more targeted, local rather than national; it promotes a continuous and gradual improvement rather than an overall questioning of the social model. Two observations can be made at this level of our development effort to better define this new form of activism for Sustainable Development 2.0.

First, we must emphasize the local and participative nature of these actions. Even if they are using the Internet to mobilize the greatest number of individuals or "consum'actors", the manifestation of this activism is not virtual but real, and is close to the ground and realities. This approach, on the one hand, echoes the major reigning principles – popular participation, local democracy and local organization; throughout the world this refers to both social movements and development organizations. It fits well into the English-Speaking movement of Community Power and Grassroots Democracy

[KAU 97]. But, secondly, via the Web, this approach attempts to circumvent the traditional obstacles encountered by traditional community participation: the power of the central administrative apparatus, lack of technical and organizational experience at local level, social divisions and the impact of national and transnational organizations. Sustainable Development 2.0 could thus deftly combine the approach of "social movements" within the reach of the Web to produce a more inclusive development model.

Secondly, we must continue our analysis by focusing our attention on the corrective and regulating extent of these private social initiatives. Without moral discourse, the purpose of these community events appears to be less focussed on challenging the operating principles of the society but rather focussed on participating in a new self-regulatory approach of the society. However, the scope of this new form of social regulation is proportional to the ability of these individually isolated individuals to come together to sensitize the society of the social or environmental problems.

Based on this observation, we can argue that the principle of Crowdsourcing – a phrase coined in 2006 by Jeff Howe and Mark Robinson, editors at Wired magazine, which reflects the idea of using the creativity, intelligence and know-how of a large number of Internet users, at lower cost, on the model of outsourcing – is a wonderful engine for self-regulation.

Sustainable Development 2.0 Websites thus participate in the construction of a new regulatory political order. They emerge in an international context where traditional regulations – especially the state law – seem ineffective in regulating certain business practices, and where companies are trying to impose only their own development model.

Our analysis of Web 2.0 is intimately linked to the issue of state regulatory crisis in the world and especially in France [MAU 07]. We here invoke the problem of the inability of the contemporary French state to establish only its authority in the society. The rule of law only is no longer effective. And this trend is not unrelated to the slowing down of the regime of a number of our institutions: "the American mythology of the rule of law is a direct outcome of legalism and puritanism of the Anglo-Saxon, as the French fascination for power is of the old Latin. Respect for the law is primarily a form of American moralism and idealism and goes hand in hand, as viewed by Tocqueville, with the respect for religion" [COH 85]. Social order is no more the result of only the state-inspired rule of law. The emergence of Activism 2.0 Websites makes us ask ourselves the following question: what are the interactions between a society's ability to change and the integration of these new considerations, praising participatory democracy and using ICT, in a society?

"The mode of regulation of a society and its idea of democracy are far from being irrelevant to its vitality and capacity for innovation and adaptation to the new global environment" [COH 85]. If one assumes that the initiative for Sustainable Development 2.0 is an expression of a form of social regulation which is superimposed on a minimum legal framework, then this mode of regulation is indicative of the ability of a business environment – the French economy and particularly its firms – to adapt to new global economic and digital environment.

These social and collaborative sites actively participate – on this general subject of Sustainable Development 2.0 – to the emergence of another model of regulation: that of self-regulation of the society. In this model, Internet users try to find their place in between the state and enterprises by establishing their own control standards. This phenomenon of emancipation is not neutral to answer the question of the developmental stage of a society. "Self-regulation reflects a greater maturity and autonomy of civil society than the state mode of regulation, where the society is under a guardian like a child" [COH 85].

3.3.2. *Web 2.0 for environmental impact*

Web 2.0 brings about a double evolution: that of use and technology. It is thus the bearer of hopes and reality in environmental matters.

3.3.2.1. *The hopes of dematerialization*

Wikipedia, a free and collaborative encyclopaedia, has provided a very interesting article on sustainable development, a chapter on the relationship between "dematerialization and sustainable development[7]". The main issues are well presented in this article and we shall thus summarize and put them into perspective in the following sections.

---

7 http://fr.wikipedia.org/wiki/Web_s%C3%A9mantique_et_d%C3%A9veloppement_durable#
Syst.C3.A8mes_d.27information.

The massive computerization of the economy over the last 50 years has moved the world into a so-called intangible economy, in which increased computing-driven management workflow has been accompanied by a parallel increase in flow of market goods and the quantities of natural resources consumed. Thus, initiatives of sustainable development principles in computing often concern hardware recycling, energy cost reduction or the reduction of paper consumption by dematerialization. Indeed, the amount of paper consumed by all transactions of the merchant world has never ceased to grow, as the case is with other raw materials.

However, dematerialization – the transformation of hard document processing into digital processing – is often presented (including by specialists in sustainable development) as a benefit with an environmental perspective, because it would decrease or even eradicate the consumption of paper and other raw materials. It is quite commonly classified as good environmental practice whenever attempts are made to cross sustainable development with Web use.

However, such an assertion is too fast: first, because a "paperless" situation does not exist, and secondly because dematerialization does not really remove the use of raw material. Information is always stored in hardware whose energy consumption is very important and the carbon impact is very real. In other words, the qualitative analysis of advantages and disadvantages of dematerialization in terms of sustainable development shows that things are not so simple. In particular, this process does not improve the real environmental quality of products. Dematerialization remains more of a modeling tool which is sometimes interesting (see the below section on virtual water) rather than an environmental solution.

Virtualization of the impossible: virtual water

"Virtual water" does not flow from any source but is increasingly feeding researchers' reflections. Participants at the Fifth World Water Forum in Istanbul in April 2009, discussed the bordering future and current realities, since currently, more than 850 million people worldwide are still deprived of drinkable water and over 2.6 billion lack sanitary facilities. Urbanization and rapid population growth would, moreover, increase pressure on this essential life resource.

Therefore, if water is now relatively stable in volume, managing it has become a major issue. The "virtualization of water" is a concept that could meet the challenge of good management of the blue gold. This conceptual approach, developed by John Anthony Allan, is now highlighted by some scientists who advocate the idea of quantifying the amount of water needed to produce goods or service. The analysis of the life cycle of the later, right from the well from which water is extracted to the finished product, would enable us to calculate its costs in market exchanges. Yet today, water needed to produce a good or a service is not included in the final price. It is sometimes more interesting to import goods to save the cost of water needed to produce it. Today, this virtualization exercise allows integrating geopolitical data into economic considerations; some countries use it to rethink their strategic directions. "In theory, countries with limited resources should export food with low water content but expensive, and import cheap products which contain lot of water" (2009).

Thus, the real challenges posed by sustainable development are related to the sharing of environmental and social information between enterprises and government and with other stakeholders, and the structuring of this information on open democratic sites, which offer an opportunity for everyone to participate in the development and policy choices. Web 2.0 makes it possible to manage this information (environmental in nature) and to structure it to implement societal networks. Sustainable Development 2.0 represents a real challenge, which can be illustrated in reflections on Mobility 2.0 (see the section below).

Mobility 2.0: soaring ride-sharing sites

In early 2009, nearly a hundred ride-sharing sites were opened for individuals and businesses, local authorities and general councils.

Ride-sharing, at the crossroads of environment, mobility, and new information technologies, is the bringing, via a Website, several people in one car to journey together. Ride-sharing tools of the new Web 2.0 enterprises enable us to reduce the number of vehicles in circulation, and thus the emission of gas relative to the green-house effect. They can increase the mobility of people without vehicles while reducing transportation costs. Their impact corresponds fully to the pillars of sustainable development since it is:

– economical, car usage is optimized according to the available five spaces, and the gas cost, tolls and parking fees are shared;

– environmental, as traffic, parking problems, pollution and energy consumption are reduced;

– supportive, as it encourages mutual help and enables those who are not necessarily owners of private cars to have access to mobility;

– social, allows people to meet one another.

In France, the concept of the first ride-sharing site "www.123envoiture.com" evolved from the following observation: throughout the country, at office closing hours, virtually all the motorists were driving alone. The site was then created with the aim of bringing drivers together to share their journeys. The site, enjoying a growing success, has more than 269,000 users, and is enriched by 400 registrations and 800 trips per day.

Their practices are now very diverse and their success continues to grow. Thus, in the Gruchet le Valasse township in Seine-Maritime, the Carrefour hypermarket uses a collaborative Website to build customer loyalty and to withstand competition. They are not the only ones employing this kind of initiative; the Casino Group has already deployed a similar system over a hundred hypermarkets. The retail sector is not the only sector using these service sharing sites, similar initiatives have been tested by the Paris-Normandy auto-routes. Even though this may mean lower toll income, it is also part of the group's strategy to reduce its carbon footprint.

These ride-sharing sites, combining social networks and positive impact on the environment, are very popular all over the world particularly in the United States. These 2.0 transport systems are provided with, in some of the states, dedicated lanes, preferential parking lots, or more toll free advantages. This is quite right, since ride-sharing is perfectly within the logic of sustainable development.

Many initiatives are proliferating in this direction in the world. It would be difficult to present all of them. For example, the Green Fund site intends to establish an ecological investment fund by demanding a hundred dollars from volunteer Internet users on social networking sites (Facebook for example) ...the fund's shareholders who will then vote for which project or which business, with a positive impact on the environment, they want their money to be used for. But all these initiatives make collaboration an obligatory point of passage [CHA 09] to encourage good environmental practices.

3.3.2.2. *The relative impact of information and communication technology on the environment*

Sustainable development and ICT has strong links which can be ignored by only those who confuse sustainable development with ecology. Indeed, the Lisbon conference has shown that knowledge is the perfect lever for trying to resolve development issues.

International institutions are constantly reminding us that the three dimensions of globalization are demography, growth and knowledge.

Similarly, the development of these technologies can help achieve the millennium goals in three key areas: poverty and hunger reduction, environmental protection, and strengthening of a global partnership for development. This contribution may be achieved through information transit.

Objective 1: Eliminate extreme poverty and hunger
Objective 2: Provide primary education for all
Objective 3: Promote gender equality and empower women
Objective 4: Reduce mortality of children who are under 5 years of age
Objective 5: Improve maternal health
Objective 6: Fight against HIV, malaria and other diseases
Objective 7: Ensuring a sustainable environment
Objective 8: Developing a global partnership for development

Table 3.1. *Millenium objectives (source: Wuppertal Institute)*

Moreover, the most polluting sectors are rather older ones. Thus, for many analysts, the challenge would be to move from an industrial economy that is a heavy consumer of raw materials and high emitter of waste to an economy based on knowledge and services. Web 2.0 is mostly at the core of these development issues.

This transformation is not easy and can lead to many adjustments, both technical and psychological or legal (see example of downloading music in the section below).

Music downloading: an environmental contribution in opposition to the copyright protection

The society must be dematerialized. To achieve this, we need to reduce the material input and increase services. Several examples of dematerialization can be cited:

– sharing hardware (study carried out with Hewlett Packard) that allows a significant reduction of the ecological backpack;

– music downloading. There are three ways to listen to music: by buying a physical CD, buying a CD online, and by downloading music.

Music downloading has the lowest impact on the environment except if the user has only a modem and not a broadband. This scenario requires the development of new policies *vis-à-vis* the consumer (broadband usage).

Telecommuting, made possible through the use of ICT, also leads to reduced pollution (less transportation) except when the leisure time of teleworkers is spent on activities that require driving.

These examples show that the union between sustainable development and new technologies has a bright future ahead, where the ecological argument remains strong.

The international organization the Wuppertal Institute, specializing on issues related to climate, environment and energy, has carried out many research works which have referenced the environmental consequences of ICT.

Three levels of distinction can be applied:

– the level of products used by ICT and the infrastructure of these products;

– the level of application of ICT (telecommuting);

– the level of related effects: changes and new habits arising from ICT.

These three levels of analysis have converged to prove that ICT emits less CO_2. An analysis of CO_2 emissions per sector carried out in a research program developed with the European Commission also shows that the ICT sector, with 0.6% of CO_2 emissions, does not seem to contribute to greenhouse gas emissions.

A nuance can yet be advanced on this environmental issue relating to ICT. If the ICT manufacturing sector is not a major contributor to CO_2 emissions, the use of ICT may not imply the same thing. ICT usage is very energy consuming. The use of the Internet could account for 6% of our energy consumption, the equivalent to the energy supplied by 3.5 nuclear power plants in some years.

All the arguments plead toward making ICT, Web 2.0 and stakeholders satisfactory tools for ensuring sustainable development.

3.4. Conclusion

Our conclusion may take the form of a borrowed transposition, of course, with humility. Tim O'Reilly's book published in 2005 under the title "What is Web 2.0?" highlighted the key principles of Web 2.0. Our exploratory work toward Sustainable Development 2.0 is expected to attempt a mapping exercise to identify key principles to answer the question: "What is Sustainable Development 2.0?":

– considering stakeholders as "co-developers" of a company's strategy and development;

– an organization's performance progresses when the differing perspectives of its stakeholders increase. Web 2.0 users have unique data which are difficult to recreate, and whose richness increases with the increasing number of users. The combination of information from sustainable development actors creates shared values;

– shared information amount to richness: all Web 2.0 applications (social, societal or environmental) are linked to a shared and specialized database;

– benefit from collective intelligence and adopt the principle that the nature of sustainable development often induce greater responsibility and awareness on social, economic and environmental issues;

– promote a continuous improvement process and establish flexible and lightweight interfaces based on new Web standards and protocols;

– if Web 2.0 is regarded as a platform for providing Web applications services to users, an organization mobilizing Sustainable Development 2.0 would be a structure that changes its business model to integrate a platform of service in place of, or alongside with, its products. Developments relating to mobility 2.0 illustrate this principle.

3.5. Bibliography

[BAR 86] BARLEY S., "Technology as an occasion for structuring: Evidence from observations of CT scanners and the social order of radiology departments", *Administrative Science Quarterly*, vol. 31, no. 1, p. 78-108, March 1986.

[BEC 01] BECK U., *La société du risque – sur la voie d'une autre modernité*, Aubier, Flammarion, Paris, 2001.

[BER 38] BERNARD C.I., *The Functions of the Executive*, Harvard University Press, Cambridge 1938.

[BLA 09] BLANCHON D., *Atlas mondial de l'eau*, Autrement, Paris, 2009.

[BOA 02] BOASSON C., *La responsabilité sociale des entreprises*, Entreprise et Personnel/CSR Europe (cahier), 2002.

[BRO 02] BRODHAG C., "Ethique d'entreprise et développement durable", *Entreprise Ethique*, (16), 2002.

[BRO 04] BRODHAG C., GONDRAN N., DELCHET K., "Du concept à la mise en œuvre du développement durable : théorie et pratique autour du guide SD 21000", *Vertigo*, 5 (2), p. 11, 2004, http://www.vertigo.uqam.ca/pdf/vertigovol5no2.pdf.

[CAP 02] CAPRON M., QUAIREL-LANOIZELEE F., *Mythes et réalités de l'entreprise responsable - Acteurs, enjeux, stratégies*, La Découverte, Paris, 2002.

[CAR 89] CAROLL A.B., *Business and Society: Ethics and Stakeholder Management*, O.H., South Western, Cincinatti, 1989.

[CAS 04] CASTELNAU P., NOEL C., "Engagement pour un développement durable et performance des entreprises : le cas français", *Journée du CERMAT sur la performance*, IAE Tours, 15 January 2004.

[CHA 09] CHAPTAL A., "Rhapsodie sur la collaboration : le travail collaboratif ", *Les dossiers de l'ingénierie éducative*, no. 65, p. 88-90, 2009.

[CHA 98] CHARREAUX G., DESBRIERES P., "Gouvernance des enterprises : valeur partenariale contre valeur actionnariale", *Finance Contrôle Stratégie*, vol. 1, no. 2, p. 57-86, June 1998.

[CHI 72] CHILD J., "Organizational structures, environment and performance: The role of strategic choice", *Sociology*, vol. 6, p. 1-22, 1972.

[COH 85] COHEN-TANUGI L., *Le droit sans l'Etat*, PUF, Paris,1985.

[DHU 05] D'HUMIERES P., *Le développement durable, le management de l'entreprise responsable*, Editions d'Organisation, Paris, 2005.

[DIG 07] DIGIMIND SERVICES, Le Web 2.0 pour la veille et la recherche d'information, white paper, June 2007.

[DRU 01] DRUCKER P., "Will the entreprise survive?", *The Economist*, 3 November 2001.

[DUB 02] DUBIGEON O., *Mettre en pratique le développement durable*, Eyrolles, Paris, 2002.

[ELK 97] ELKINGTON J., *Cannibals with Forks, The Triple Bottom Line of 21st Century Business*, Capstone Publishing, Oxford, 1997.

[FER 01] FERONE G., D'ARCIMOLES C., BELLO P., SASSENOU N., *Développement durable,* Editions d'Organisation, Paris, 2001.

[FRE 84] FREEMAN E.R., *Strategic Management: A Stakeholder Approach*, Pitman Publishing, Boston, 1984.

[FRI 02] FRIEDMAN L., MILES S., "Developing stakeholder theory", *Journal of Management Study*, vol. 39, no. 1, p. 1-21, 2002.

[GEN 03] GENDRON C., LAPOINTE A., TURCOTTE M.F., "Codes de conduite et entreprise mondialisée : Quelle responsabilité sociale ? Quelle régulation ?", *Les Cahiers de la chaire économie et humanisme* (UQMA-ESG), vol. 12, p. 22, 2003.

[HAB 77] HABERMAS J., *Droit de la démocratie*, Gallimard, Paris, 1977.

[HEN 90] HENDERSON R., CLARK K.B., "Architectural innovation: The reconfiguration of existing product technologies and the failure of established firms", *Administrative Science Quaterly*, vol. 35, p. 9-30, 1990.

[IGA 02] IGALENS J., JORAS M., *La responsabilité sociale des entreprises : comprendre, rédiger le rapport de développement durable*, Eyrolles, Paris, 2002.

[KAU 97] KAUFMAN M., DILLA ALFONSO H., *Community Power and Grassroots Democraty, The Transformation of Social Life*, CRDI/Zed, 1997.

[LAI 05] LAIDI Z., *La société civile internationale existe-t-elle? Défaillance et potentialités, Culture générale*, t. 2, PUF, Paris, 2005.

[LAV 02] LAVILLE E., *L'entreprise verte. Le développement durable change l'entreprise pour changer le monde*, Village Mondial, Paris, 2002.

[MAR 58] MARCH J.C., SIMON H.A., *Organizations*, John Wiley & Sons, New York, 1958.

[MAU 05] MAULEON F., "Introduction au Management durable", in D. WOLFF and F. MAULÉON (eds), *Le management durable*, Hermès, Paris, 2005.

[MAU 07] MAULEON F., La communication extra financière comme expression de l'éthique de l'entreprise, PhD thesis, Toulon, 2007.

[MAU 10] MAULEON F., GIOANI M., Welcome to CSR 2.0: Networking firms, managers and citizens to face CSR challenges, Communication in Université d'été de l'IAS, Pau 2010.

[MER 99] MERCIER S., *L'éthique dans les entreprises*, La Découverte, collection Repères, Paris, 1999.

[PER 67] PERROW C., "A framework for the comparative analysis of organizations", *American Sociological Review*, p. 194-208, April 1967.

[PER 02] PERSAIS E., "L'excellence durable : vers une intégration des parties prenantes", *11e Conférence de l'Association Internationale de Management Stratégique*, Paris, 5-7, June 2002.

[PER 03] PERSAIS E., "Le rapport de développement durable (ou Stakeholders' Report). Un outil pour une gouvernance sociétale de l'entreprise?", *Atelier de l'AIMS "Développement durable"*, Angers, 15 May 2003.

[PLA 03] PLANE J.M., *Théorie des organisations*, collection Les Topos, Dunod, Paris, 2003.

[POW 96] POWELL W., KOPUT K., SMITH-DOERR L., "Interorganizational collaboration and the locus of innovation: Networks of learning in biotechnology", *Administrative Science Quaterly*, vol. 41, p. 116-145, 1996.

[OCD 01] OCDE, *Responsabilité des entreprises. Initiatives privées et objectifs publics*, Editions de l'OCDE, Paris, 2001.

[SPI 99] SPICHER P., "Lien entre performance éthique et performance financière : vers un consensus ?", *Entreprise Ethique*, no. 11, p. 39-44, October 1999.

[TAY 11] TAYLOR F., *The Principles of Scientific Management*, Harper & Row, New York, 1911.

[THO 67] THOMPSON J.D., *Organizations in Action*, McGraw-Hill, New York, 1967.

[TUS 90] TUSHMAN M.L., NELSON R.R., "Introduction: technology, organizations, and innovation", *Administrative Science Quaterly*, vol. 35, p. 1-8, 1990.

[VIL 98] VILLENEUVE C., *Qui a peur de l'an 2000 ? Guide d'éducation relative à l'environnement pour le développement durable*, Editions Multimondes et UNESCO, Sainte-Foy, 1998.

Chapter 4

Corporate Education and Web 2.0

4.1. Introduction: what is corporate education?

Corporate Education (CE), which is normally associated with the concept of Corporate University (CU), has emerged as a reality that surpasses our imagination when we seriously think about this field. Historically, we can say that corporate universities were born out of the evolution of training activities proposed by companies' human resource departments. Initially, their principal objective was "…to prepare their employees toward the improvement of their professional competence and excellence in their activities within the company" [TAR 09].

With the present evolution of the scope in corporate universities, the concept of the corporate university appears to be one of the best ways for companies to educate their employees, in order to ensure their competitiveness in the market. We will share the views of authors like Allen. For him, a CU is an educational entity which is a strategic tool. It is designed to assist the organization to fulfill its mission by accomodating those activities that create a culture of individual and organizational learning, leading to knowledge and wisdom. For Allen, the word strategic is the more important one in this definition [ALL 02].

Chapter written by Miguel Rombert Trigo, Alice Maria Salgado Gonçalves and João Casqueira Cardoso.

Eboli [EBO 2004] contributes by complementing this definition by stating that the purpose of corporate education is "...the development and installation of entrepreneurial and human skills considered critical for business strategies".

Rosa [ROS 2008][1] points out that while professional training is geared toward learning specific skills to perform tasks, corporate education focuses on a higher level of people's intellect by preparing them for life and work. As stated in his defence of CE, irrespective of the strategic itinerary of the organization, "the new frontier of competitiveness will place more demand on knowledge, innovation, and human capital".

Analyzing the motivations behind the establishment of corporate universities by various organizations, we can conclude that the most relevant reasons are the following:

– immediate improvement of employees' performance in carrying out their functions;

– a greater commitment between employees and the organization through a more elaborate spread of institutional values and culture;

– dynamization of higher dimension leadership programs (with respect to learned skills and the number of participants);

– creation of a catalyst that helps an organization achieve the necessary organizational changes;

– financial issue: companies certainly want to invest in courses that provide guarantees for the acquisition of knowledge which will translate into real gains for their business.

It is certain that the motivation, by itself, is not sufficient, and several researchers have carried out some works in order to identify good practices which can help ensure that CU projects would guarantee success. Meister [MEI 98] highlights the essential elements that must be in place, so that a corporate university could succeed:

---

1 Member of the Accor academy (Brazil) founded in 1992 – a period in which the country began to project itself internationally and its companies prepared an ambitious program to develop their businesses.

– linking the development objectives of the training for the organization's strategic needs;

– involving the leaders as students and teachers;

– electing an executive responsible for organizational learning – named "Chief Learning Officer";

– considering the training of employees as a continuous strategic process and not as a one-time event;

– linking employee compensation to knowledge acquisition, sharing, and implementation;

– widening the scope of action of the corporate university beyond the training of company employees. Ensuring that the training targets the customers as well as the suppliers;

– making the corporate university function as a center oriented toward business within the company itself;

– developing a series of new innovative alliances/partnerships with higher education institutions;

– promoting the value of the learning infrastructure of corporate university;

– developing the corporate university as a tool for competitive advantages and as a business center.

Considering these declarations as a reference, we recommend that corporate education can and must play a broader role, where corporate education functions as a fundamental pillar (with strategic innovation, competitive intelligence, continuous improvement and the promotion of creativity) for the organizational continuity throughout the 21st Century. We are convinced that corporate education should be considered as the best model for the education of working adults.

In fact, adults are facing growing challenges and problems of increasing complexity while performing their duties. Therefore, corporate education should become a synonym of adult education programs – programs that play or should play an active role in the business world by focusing on the needs for practitioners to acquire and develop new skills, by increasing their employability conditions and consequently leading to improvements in their companies' performance. Corporate education must result from strategically delineated programs, which link the best of applied research to commensurate experience through knowledge gained from practical experience in corporate education.

4.2. Evolution of corporate education

We can state that the genesis of corporate education can be traced to the creation of the initial corporation schools in American territory. It emerged on the industrial scene, between the second half of the 19th Century and the first half of the 20th Century, in a context where the production process has been designed from an engineering perspective.

> "Corporation schools began to emerge during the mid-19th Century in USA, when large companies such as DuPont and Edison set up technical and liberal arts education for current and prospective employees... were intended to correct perceived inadequacies in state provision, as employers reportedly found it difficult to recruit employees with the requisite skills or attitudes" [STO 04].

The main objective of training was to enable people to perform tasks according to available models and time in a real economic and mechanical interpretation of work and management theory, championed by men like Frederick W. Taylor – theories that reduced human beings to their role of *homo economicus* [FER 01]. However, the initial decades of the 20th Century were also influenced by the new works related to the social sciences and humanities fields, especially with the results of a study developed between 1924 and 1927 by Elton Mayo and his team.

These researchers had introduced the discussion on the importance of informal groups in socialization and cooperation. They testified that the organizations formed by human beings seek recognition, harmonious interaction, participation, decision-making power and job satisfaction. Following this conception, human beings started to be characterized, not only as being driven by economic objectives, but as a *homo sociologicus* and a *homo psicologicu*s [FER 01].

The years following World War II were characterized by rapid economic, social, political, technological, cultural changes and an accelerated economic growth. These events led to enormous changes in organizations and consequently, in ways to manage the staff. Organizations became more democratic, they began to dedicate more value to their human resources, an essential factor for competitiveness, by creating models with an emphasis on people's participation in decision-making [FER 01].

It is within this context, according to Tarapanoff, that the first corporate university created at a global level was born, more precisely in 1945 in Crotonville (New York). It was General Electric's corporate university and, in its first version, it was nothing but a deepening of the Human Resources Department by offering courses for managers to improve their activities. The disclosure of the initial results obtained by General Electric, linked to the development of its corporate school, had made the Crotonville Learning Center become a success case study and enabled the concept to gradually gain admirers and supporters.

During the 1950s and 1960s, by virtue of the competitiveness and the acceleration of technological evolution, knowledge and updated learning became essential. To survive, organizations needed to possess the ability to learn and innovate at a faster pace. Tarapanoff indicates that the major economic groups in those days faced enormous challenges. One of the most complex those challenges was the effectiveness in management to respond to situations such as the opening of new offices (many of them overseas) and the accelerated recruitment of staff, knowing that it was necessary to maintain and transmit organizational culture and values in parallel. It is within this context that it was found in the 1970s that people's professional growth process cannot only be based on training (for improving how a function or position is performed), but also on the development of individual capacities.

The 1980s were marked as a period of significant changes, especially with the booming technology market, in which organizations were confronted with increasing challenges in terms of their survival skills. Most of the innovative companies adopted structures that were more flexible, more decentralized, and simpler in order to create a space for creativity, autonomy and accountability of employees.

Research and development centers had begun to proliferate. Managers had to find quick formulas for the development and motivation of their human resources, resulting in value employees feeling valued, and as a consequence employees developed innovative ideas. We were entering the era of *homo educandus*, a person in constant learning, because as stated by Senge [HEN 97], the "only source that provides sustainable competitive advantages, is to learn faster than one's competitors".

At that time, corporate education projects started to record success, with the Motorola University taking the lead. This structure acted as a partner who was oriented toward the parent company's markets, seeking solutions for talent development, knowledge acquisition and raising the organization's intellectual capital, as described by Torres [TOR 09].

Today, through the positive results that corporate education projects have been able to achieve, the existence of corporate universities has become a global phenomenon. In the United States, where there are over two thousand corporate universities, these structures have become realities in big enterprises like Toyota, Sun, Motorola, McDonald's, or IBM [VAL 07].

The present CUs have little to do with the first of its kind which destined the training to an audience of executives at a purely internal level. Today, corporate universities include all employees of the organization, and corporate education is extended to all stakeholders of the organization.

Meister argues that the traditional model of the corporate university is changing: moving away from the model of a cost center to a unit that is self-financing (by charging for its services), and may therefore become a source of profit for the parent enterprise.

At present, administrations want their corporate universities to resemble higher education institutions by developing a broader set of training programs and charging the internal clients of the organization[2] for the training. The materials used in the corporate university's programs should be of high class, so that they can be marketed to consumers and suppliers.

We argue that the implementation of CE in an organization is a holistic project, and that it is reasonable to develop it, if and only if the entire organization is truly committed to it, so long as results of people's commitment and development are visible. Following this perspective, new business models that are more appropriate for each situation will emerge between people and organizations.

---

2 These clients can be, for example, business directors or production managers.

4.3. Corporate Education 2.0

Given the fact that the 2.0 phenomenon is a concept that involves paradigm shifts in all organizations who claim to have a say over their destiny, we argue that the use of corporate education should be a stimulator of cultural and organizational transformations. We also believe that CE will guarantee a more complete lifelong learning path for people, who are becoming increasingly more important for the implementation of differentiating strategies that ensure competitive advantages. Organizations in tune with their time perceive that it is more important than ever to think of human capital education as one of the pillars of organizational strategy and a way to ensure their survival and sustained growth.

In our view, Corporate Education 2.0 (CE 2.0) must express itself clearly as an important ally of organizations. CE 2.0 units must play its role proactively, and must be capable of generating businesses in the interest of the entire organization. It must therefore be an unit of excellence in the services it offers so that it can succeed in transforming the investment into results (in direct form, through the services and products that it negotiates and in indirect form, through the results that will be achieved by teams that it has formed). To achieve this result, we are convinced that the Corporate Education 2.0 must be based on at least two main pillars:

– Andragogy[3] models: this entails finding and implementing corporate education actions in the best environments and contexts which can ensure exchanging and sharing of knowledge. It is interesting to note that despite the existing consensus that corporate education must be based on the best experiences for their applicability in professional practice, developed projects often seem to depend on traditional models used in the most conservative business schools. This finding is also corroborated by May: "the majority of corporate universities copy the model of academic institutions both in the structure and in the objective. For the majority of corporate universities, the goal is to promote education, and they become beautiful central training department" [MAY 06];

– Education as a global vision vector: since this chapter is in context of a book on competitive intelligence, we cannot cease to consider that CE 2.0 must strategically rely on systematic analysis of the 21st Century with

---

3 Andragogy is adult pedagogy, the term comes from the Greek *anêr* (from *andros*) "adult man" and *agô* "lead, drive, carry, raise".

respect to trends and changes in the competitive environment. We will use this reflection, as well as those outlined in the chapter on Management 2.0, to suggest how CE 2.0 can respond to new demands, opportunities and threats faced by organizations.

This situation implies rethinking the role of employees in organizations. The era of knowledge workers, (a definition that was used by Drucker [DRU 56]), is now giving way to the conceptual age PIN [07]. The latter is characterized primarily by the need of these workers to rethink the way they consider work, responsibilities and participation in organizational life.

"Many of the current knowledge professionals... should possess a whole new set of skills. They should do, with the same quality and with much less money, what foreign employees fail to do..., build relationships instead of transactions by attacking new challenges instead of deciding routine problems, and synthesize several elements in opposition to analyzing a single element each time" [PIN 07][4].

We can therefore argue that the first task that must be considered while realizing CE 2.0 is to help people to initiate processes of personal re-engineering by developing new soft skills[5] which enable them to transform their weaknesses into strengths and develop their ability to discover new opportunities for themselves and their organizations. The formative curriculum should address the following topics: trends and technological innovations; social, economical, demographical and consumer trends, and transformations in the competitive environment.

4.3.1. *Trends, technological innovations and Corporate Education 2.0*

In terms of technological trends, it is easy to recognize the frenzied pace at which information and communication technologies (ICT) cease to be at the forefront and become obsolete, replaced by others that are more advanced and with a higher level of efficiency. According to the available historical data, the useful lifetime of technological innovations in this regard tends to be shortened.

---

4 Extract translated by the authors of this chapter.

5 For example: adaptation capacity, team work capacity, empathy.

1700 generations ago modern "man" emerges and starts to develop language

300 generations ago, writing was invented

35 generations ago,
Printing press was developed

and then…

2000	Cellular/Web
1985	PC/Networking
1970	Fax/Electronic
1955	Television/MassMedia
1940	Radio/Talkies
1925	Telephone/silent films
1910	Telegraph/Photography

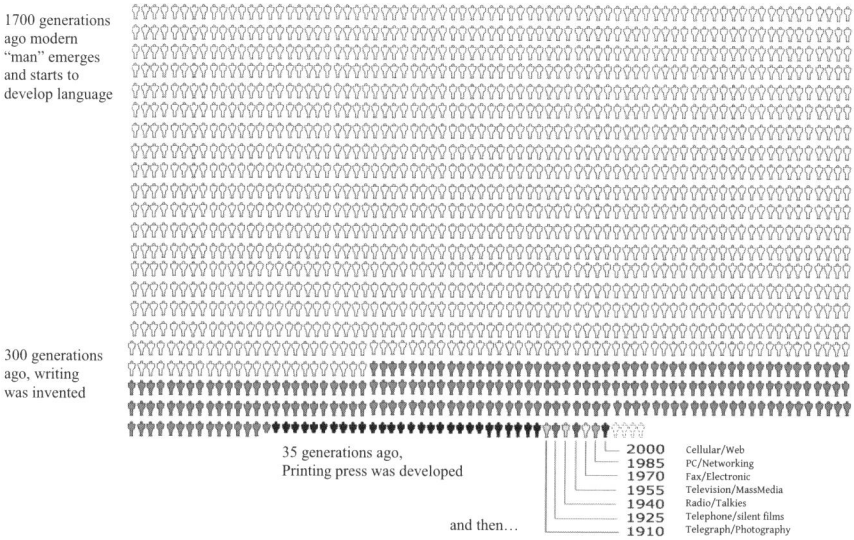

Figure 4.1. *Technological evolution according to René Barsalo [BAR 04]*

Advances in information and communication technology undoubtedly provide improvements in the way we work and communicate. However, this phenomenon also tends to confront people and organizations with new realities.

As argued by Pink, human beings are the best option for many tasks, "but it is important to be aware that to achieve many activities that depend on a logic based on rules, calculations and sequential reasoning, and computers are simply better, faster and more competent" [PIN 07][6].

The biggest technological breakthrough that has occurred in recent decades is the Internet phenomenon. The Internet has enabled a communicative and entrepreneurial paradigm shift by transforming the way we visualize business models since, as we all know, the Internet facilitates transparency and exchange of data, and improves efficacy by facilitating mobility.

The function of CE 2.0 should be to identify the best ways people can use technology for their own good, and help to equip them with new skills and

---

6 Extract translated by the authors of this chapter.

knowledge which will allow a smooth transition to the new model that helps to realize daily tasks.

In summary, success in using Web 2.0 tools, which are made available through the Internet to employees of the 21st Century, will enable the transition to Enterprise 2.0[7]. These practices can be taught. In fact, employees and organizations have to acquire soft skills, both at individual and collective levels. This transition will also result in productivity gains, particularly in finding new market niches that would not be possible without using new technologies.

4.3.2. *Social trends and Corporate Education 2.0*

In Chapter 2 entitled "Management 2.0", we saw the importance that the Net generation will have in our society in the near future. CE has an important role in the development of educational programs that enable organizations to prepare their employees to perceive the more significant changes in terms of standards and values.

It is absolutely necessary to prepare the company staff, accustomed to a certain model of organization and execution of work, to cohabit with members of the Netgeneration and their eight standard features[8]. They represent a new class of employees entering the workplace. They also represent a new class of consumers, with ambitions and needs which should be satisfied in terms of products and services. It is a fundamentally new world and Corporate Education 2.0 should teach the staff to associate with this new generation.

The role of CE 2.0 is to work toward changing the paradigms that are rooted for a long time in an institutional organization. On one hand, this effort would help every member of the organization to understand how to incorporate the necessary changes, and how to empower them to achieve the best results. On the other hand, this effort would also enable the organization to understand that the young ones (Net generation), as new consumers, would like to behave in a far less passive way than the previous generations.

---

7 "Enterprise 2.0" *Enterprise 2.0 is the use of emergent social software platforms within companies, or between companies and their partners or customers* [MAC 11].
8 Freedom, Costumization, Scrutiny, Integrity, Collaboration, Entertainment, Speed And Innovation [TAP 08].

We propose that the training programs like "knowing the Net Generation", opened to all company departments, aim at understanding the eight standards and the changes that these standards are likely to generate in the company. Such training programs should be conducted on two important points and address the following content:

– net generation at workplaces:

   - collaborative work,

   - from recruiting to motivation: rethinking how we recruit, reward, develop and supervise talents;

– net generation in the market:

   - knowing the "prosumer",

   - co-innovation and co-creation of products and services.

4.3.3. *Transformations of the competitive environment and Corporate Education 2.0*

The Current 2.0 world comes with new trends and new transformations of the competitive environment of firms. It is essential that institutions and individuals learn to position themselves with respect to these new perspectives. Everything goes very fast or, as Hamel and Breen put it [HAM 07], "nothing goes faster than change". We must therefore directly confront these new challenges and initiate projects that can help us face them with the best chance of success.

As we indicated in Chapter 2 on "Management 2.0", the current competitive environment requires companies to incorporate innovation as a transversal competence, which should be part of the organizational culture. They increasingly need people who are in harmony with our time and who feel concerned for the company. Such people are likely to respond with creativity and elegance to the challenges faced. This way of responding to the challenges must become as natural as breathing, and in the way each employee acts.

It is clear that this posture does not come automatically or by chance. Rather, it depends heavily on a series of steps which the company must implement being the strategic education of the company's human capital one of the most important ones. CE activities can allow the emergence of a reflection space which can provide answers to the new challenges affecting the organization. It is through Corporate Education 2.0 that, as stated by Skarzynski and Gibson [SKA 08], people "can understand how to find new strategic perceptions that challenge the orthodoxy of the sector, while assimilating all that is changing in the world and finding new opportunities in this changing world, by the emergence of new forms of skills and actions, and growth of a deep understanding of customers' needs which are not always explicit" [SKA 08].

4.4. Good examples of Corporate Education 2.0

We do not pretend to present all the great case studies of corporate education since it will not even be possible to present all of them in a sub-section of a book chapter. However we will be presenting a few good examples which we have come across during our research work.

4.4.1. *Toyota University: Lean learning*

The Toyota University was created in the United States in 1998 at the initiative of Toyota Motor Sales USA Inc., with headquarters in Torrente (California), with the recommendations of the strategic business plan – "Strategy of the new era" – to improve Toyota's competitive advantages.

Rooted in their corporate culture, managers responsible for planning the Toyota University embarked on an exhaustive search of existing best practices in corporate education. Not satisfied with what they found, they considered that the corporate university should be based on two fundamental values of the firm: continuous improvement and respect for people.

From a report written by May, we could identify some best practices that allow us to present Toyota University as a good example of CE 2.0.

"Toyota University consists of… a study on working with lean technical knowledge[9]… an integrated work, entirely focused on the creation of intellectual capital and building a profound discipline on the solution of practical business problems" [MAY 06].

---

9 Lean is centered on preserving value with less work.

This author also adds that the goal was to successfully develop "a model that would serve as a lever to ensure mastery in the field of production: managers as models, tutors and instructors, and practical learning, ...a chain of intelligent refueling, concieved with the same expertise as the manufacturing of Toyota vehicles... and which should be managed just like any other value chain... information, and expertise should be as fluid as parts on the assembly line of Toyota production system... a model that would add value by advancing business, more like an advisory council than like an academic institution" [MAY 06].

According to May [MAY 06], through this philosophy, Toyota University has succeeded in achieving the objective of creating "a unified strategy consisting of three targets: personal mastery, operational excellence and cutomer focus". Structurally, it has succeeded in penetrating the entire organization,

4.4.2. *GE, Leadership, Innovation and Growth (LIG)*

This company, which has been a CE global pioneer, creating the Crotonville Learning Center, continues to demonstrate an exemplary case of using corporate education as a strategic tool for organizations. After renaming its corporate university as "John F. Welch Leadership Development Center at Crotonville", through its CEO (Jeffrey R. Immelt), GE has introduced[10] the new corporate education project of the company, called LIG (Leadership, Innovation and Growth).

LIG's main objective is to support the growth strategy of the company, which, as suggested by Immelt, must be based on business development or the creation of new business, and not on buy-outs. LIG is also innovative about the adopted adult education model. As argued by Immelt, corporate education should be based on team learning, that is the reason why for the first time they had gathered together all the team members from the different business units.

But what is LIG after all? Prokesch [PRO 09] argues that the main characteristic of LIG is related to its coverage. Generally, development programs at the management level, focused on teaching and inspiring peoples' actions, face a big problem – the fact that other members of the

---

10 In a letter to its investors.

team did not attend the course may constitute a barrier to change. The author argues that LIG is the antidote to this problem in conducting the same training for the full teams. It is thus possible, at the end of CE programs, that all the managers share the same consensual vision in terms of opportunities and problems and also on how best to respond.

What are the key results of GE? Since a proactive discussion on topics is often not adequately carried out, it is the critical and in-depth analysis of issues such as those listed below that will constitute the most valued and the most important contribution of LIG to GE:

– participants are encouraged to address issues relating to the barriers to change, whether heavier barriers (organizational structure, capabilities and resources) or lighter ones (how do members of the management team behave, individually or collectively, and what their schedule is);

– the eternal challenge of management to successfully separate between the short and long term. In other words, manage the present and prepare for the future;

– to prevent GE employees from interpreting the new guidelines in a different way, courses developed a vocabulary of change, which reflects in a range of terms and words used in everyday internal and external communication of GE;

– the program is not a mere academic exercise. It was structured, so that each team can submit a draft of an action plan for establishing changes in the business and the way to go about effectively implementing these changes.

Summarily, through LIG, and as assured by Prokesch [PRO 09], GE can ensure that changes take place in a faster and more efficient way.

4.4.3. *Enclos university: develop a corporate education project for an SME*

The Enclos Corporation project, lesser known as compared to the previous projects, is an enterprise of less than 500 people with regional offices distributed all over the world. Despite its size, Enclos Corp. is a major global operator. Such a position implies greater needs in terms of employees' quality. Due to this requirement the company had decided to create a corporate university, with the following objectives:

– to enable the growth of Enclos Corp. without risking excessive speed in incorporating new employees;

– to build a culture of excellence and learning;

– to reduce risks and falls in profit margins;

– to increase efficiency through a better reusage of resources;

– to increase innovation;

– to improve a workforce that is more and more decentralized and geographically dispersed;

– to increase Enclos competitive options through knowledge, and learning.

According to Allen [ALL 07], with these fixed objectives and a team of two people, Enclos University has distinguished itself from others through the following characteristics:

– the project leader was chosen from project managers and recognized by his peers, even when he had no experience in corporate education;

– in order to develop a project that would build on best practices in corporate education, the project team made several "benchmarking visits" to corporate universities and participated in various events in this area before creating Enclos University;

– the elaboration and development of a business plan for the Enclos University was strategically focused on the principle that teaching, researching and knowledge dissemination tasks should be primarily assigned to company employees;

– definition of four action domains that should contribute to the survival of the company:

- Learning Management: although most of the training programs are conducted in the presence of those concerned, various small modules are processed live on the Web or recorded, allowing employees to have access to them whenever the need arises,

- Knowledge Management: through concrete measures make it possible to disseminate knowledge, information and best practices outside the context of formal models,

- Wisdom Management: creating a model that promotes job rotation, in agreement with formal education programs in areas considered as the most critical for each job,

- Innovation Management (R&D): innovation must find its place within the corporate university. Taking inspiration from the happenings in higher education institutions, the Enclos Corp. decided to integrate R&D domain among the fields of its university, with a mission to help ensure that the most innovative projects can provide better conditions for achieving success.

4.5. Competitive Intelligence and Corporate Education 2.0

According to Quoniam and Queyras [QUE 06, QUO 06], Competitive Intelligence (CI) is an operational system for collecting, processing and delivering tacit and explicit information to strategic decision-makers. The goal of competitive intelligence is to provide the right information at the right time and in the right way to the right person, so that he/she can make the right decision.

In other words, competitive intelligence generates knowledge for strategic decision-making, since information provided by the CI team enables taking actions in real time, and also continuous learning about competitors, customers, suppliers, and important market developments. As stated by Fuld [FUL 06], competitive intelligence can and should be taught, since it is the task of all employees of the organization.

Module	Audience	Tutors	Objectives
CI and its importance to organization	All employees	Top Management, experts in CI	Introduction to CI, participation models in CI process
Our CI project	All high level management members	High level managers/ specialists in CI	Development of CI project for organization
Seeking new opportunities	All high level management members	High level managers/ specialists in CI	Making CI become an expert in organizational competence

Table 4.1. *CE 2.0 training, applying CI to our organization*

As we have already seen earlier in this chapter, it is essential that organizations and their staff be very knowledgeable about their competitive environment, so that they can apply their skills to take advantage of opportunities offered by the market. This desideratum can be achieved by making competitive intelligence education a stake to all employees. Considering the fundamental role that competitive intelligence is positioned toward the competitiveness of organizations, we believe that it is important that training on competitive intelligence should be offered in corporate universities – according to the schema shown in Table 4.1.

4.6. Conclusion

The financial crisis that evolved in 2008 has cast a shadow of a deeper crisis (both economical and social). It is impossible to return to the competitive environment that preceded this crisis, because many companies, scalded by the events of 2008/2009, are now relapsing in the development of their projects. One concept seems to be shared in the business world: the only thing that is certain is that everything is changing.

However, it is reasonable to assume that the direct and obvious association between education and performance improvement will continue– whether at an individual, collective or organizational level. It is therefore not permissible to neglect the issue of investment in corporate education.

It is essential to invest in corporate education as a strategic asset. The training programs offered should not be mere "technical" instruments, but rather the engine, where the engine would enable the organization to attune to the world that surrounds it.

4.7. Bibliography

[ALL 02] ALLEN M., *Corporate University Handbook: Designing, Managing, and Growing a Successful Program*, Amacon, New York, 2002.

[ALL 07] ALLEN M., *The Next Generation of Corporate Universities: Innovative Approaches for Developing People and Expanding Organizational Capabilities*, John Wiley & Sons, San Francisco, 2007.

[BAR 04] BARSALO R., Société des arts technologiques, online, ref. Of Montreal April 5, 2009, available at: http://www.inm.qc.ca/pdf/midicitoyen/presentation_barsalo.pdf.

[DRU 56] DRUCKER P., *Landmarks of Tomorrow*, Harper, New York, 1959.

[EBO 04] EBOLI M., "Educação Corporativa", *Revista T&D – Inteligência Corporativa*, vol. 137, no. 12, São Paulo, November 2004.

[FER 01] FERREIRA J.M., NEVES J., CAETANO A., *Manual de Psicossociologia das Organizações*, McGraw-Hill, Lisbon, 2001.

[FUL 06] FULD L., *The Secret Language of Competitive Intelligence*, Crown Business, New York, 2006.

[HAM 07] HAMEL G., BREEN B., *The Future of Management*, Harvard Business School Press, Boston, 2007.

[HEN 97] HENRIQUES M., *Capital Humano*, p. 63, Vida Económica, Porto, 1997.

[MAC 11] MACAFEE A. Entreprise 2.0, online, ref. of March 28, 2011, available at http://andrewmcafee.org/2006/05/enterprise_20_version_20/.

[MAY 06] MAY M., *The Elegant Solution: Toyota's Formula for Mastering Innovation*, Free Press, New York, 2006.

[MEI 98] Meister, J., *Corporate Universities: Lessons in Building a World-class Work Force*, McGraw-Hill, New York, 1998.

[MEI 08] MEISTER J., Universidades Corporativas: A resposta das empresas ao ensino tradicional, online, ref. of April 5, 2008, available at http://www.tiadro.com/News/artigos/univcoorp2.html.

[PIN 07] PINK D., *O cérebro do futuro : a revolução do lado direito do cérebro*, Elsevier, Rio de Janeiro, 2007.

[ROS 08] ROSA L., "Educação Corporativa orientada ao desenvolvimento das pessoas e dos negócios", in VALIUKENAS C., DUARTE V. (eds.), *Educação faz a diferença*, Acadèmie Accor Latin América, São Paulo, 2008.

[SKA 08] SKARZYNSKI P., GIBSON R., *Innovation to the Core*, Harvard Business School Publishing, Boston, 2008.

[STO 04] STOREY J., *Leadership in Organizations: Current Issues and Key Trends*, Routledge Taylor & Francis Group, London, New York, 2004.

[PRO 09] PROKESCH S., "How GE teaches teams to lead change", *Harvard Business Review*, vol. 87, no. 1, p. 99-106, 2009.

[QUE 06] QUEYRAS J., QUONIAM L., "Inteligência Competitiva", in TARAPANOFF K. (ed.), *Inteligência, Informação e Conhecimento*, IBICT & UNESCO, Brasília, Brazil, 2006.

[QUO 06] QUONIAM L., Definitions and Concepts, online, ref. 15 December 2006, available at http://quoniam.univ-tln.fr/supports.shtml.

[TAR 09] TARAPANOFF K., Panorama da Educação Corporativa no Contexto Internacional, online, ref. of March 4, 2009, available at http://www.educor. desenvolvimento.gov.br/conhecimento.html.

[TAP 08] TAPSCOTT D., *Grown Up Digital*, McGraw-Hill, New York, 2008.

[TAR 09] TARAPANOFF K., Responsabilidade Social das empresas e a Educação Corporativa, online, ref. of April 5, 2009, available at http://www.educor. desenvolvimento.gov.br/conhecimento.html.

[TOR 09] TORRES D., Universdade corporativa: a experiência da Motorola, online, ref. of March 21, 2009, available at http://www.educor.desenvolvimento.gov.br/ conhecimento.html.

[VAL 07] VALENTE C., MATTAR J., *Second Life e Web 2.0 na Educação - O potencial revolucionário das novas tecnologias*, Novatec, São Paulo, 2007.

Chapter 5

Marketing 2.0

5.1. Introduction

Today, e-marketing is undergoing a transformation due to Web 2.0 which has changed the practice, customs, and ways of apprehending and promoting on the Web. Faced with this phenomenon, e-marketing has evolved based on new disciplines for understanding Internet user behavior, measuring accurately the effectiveness of promotional activities and increasing efficiency of Website performance. It is the birth of Web Analytics.

This new discipline is based on methods and tools consisting of many key performance indicators which provide an effective assistance for steering e-marketing.

The evolution of methods and tools is gradually converging toward competitive intelligence solutions through Web Analytics 2.0.

5.2. E-marketing: a changing activity

In 2001, Dubois and Vernette proposed the following e-marketing definition: "marketing mobilization of all aspects of the technological potential offered by new technologies for a new approach to enterprise markets". This definition of that area shows an application of traditional

---

Chapter written by Sébastien BRUYÈRE.

marketing to an Internet channel with an integration of a redundant cycle to promote e-business.

In 2006, 5 years after the initial definition, Maurizio Goetz [MOR 06], in turn, defined e-marketing as "the set of strategies, activities, and techniques for:

– optimizing a marketing information system by using acquired information in real time;

– identifying and satisfying consumers' needs at the moment, if such needs arise, through the development of a direct, interactive, personalized and long-term relationship".

This new definition evokes new elements arising from the advent of Web 2.0 and quite rightly allows making e-marketing a stand-alone discipline with its specificities, approaches, models, etc.

Thus, the 4P rule (Product, Price, Place, and Promotion) from the traditional marketing-mix formulated by Kotler, so far adapted to the Internet [LAN 07], is evolving toward a "Mix e-marketing", with the introduction of the fifth P [RIC 08]:

– Price: proposed rates on the Internet are the same or even less than those charged on other channels. The emergence of e-commerce is a strong lever for cost reduction and engaging in the implementation of attractive Internet rates to sell tickets online.

– Place: distribution truly depends on the marketing strategy intended by the company. The Internet channel can be used to generate traffic to traditional retail outlets, call centers, or instead be dedicated to the Internet. Marketing Buzz and e-merchandizing can actually be applied in this context.

– Promotion: visibility is most important factor and e-marketing entirely works in this direction. The design of a Website is often strategic to business because decision makers must provide or validate the strategic content desired to be visible to all, and as such e-marketing should be really strategic. In fact, promotion will help increase the visibility of Websites by generating more traffic and therefore, more business. E-marketing levers are numerous (SEO, sponsored links, e-mailing, etc.), and setting goals is imperative for a better choice.

– Product: products should also integrate the Internet dimension. Guarantee management, forwarding, customer service, and appropriateness with the core of business activity are attributes to be considered before posting a product on the Internet.

– Participation (Richard Lanneyrie, 2008): an emerging component of Web 2.0 is that it changes the vision of the consumer to a "responsible consum'actor" who can give his view through the provision of feedback or evaluation on items, via the Internet forum initiative. The "consum'actor" also has at his disposal price comparison services to help him during decision-making. From these events, the e-Marketers need to understand Web 2.0 for effective marketing[1].

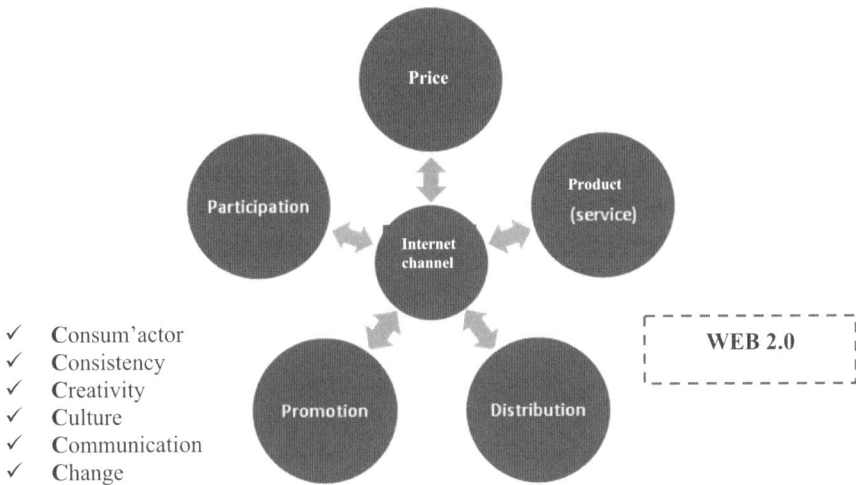

Figure 5.1. *E-marketing transformation in Web 2.0 era*

In this context, the 6C rule [BRI 06] complements the vision of the P in participation and of the P in promotion, by bringing supplements arising from the Web 2.0 revolution:

– "Consum'actor": e-marketing strategy must be focused on consumers. We must know the interests and consumer behavior on the Internet before acting on the brand, a product or a service published on the Web. Web

---

1 The virtual community of interest "is a powerful strategic lever which can unite a group of identified individuals, and subsequently would enable us to develop packages of services and products, so as to satisfy the needs of this community" [GON 07].

analytics tools, if applied with a method and objectives, would enable us to discover these behaviors.

– Consistency: this is achieved by IMC (Integrated Marketing Communication) and can act on the brand image to maximize its adequacy with the market and consumer expectations.

– Creativity: this allows demarcation of attraction. This also helps to inform, persuade, and promote memory.

– Culture: interculturalism is an important factor to be considered. A Website is accessible to the whole world and the marketing approach may be interpreted differently. In e-marketing, location-based indicators and geographic targeting would enable us to integrate this key factor.

– Communication: communication is like participation. This is a key element of Web 2.0. The Internet consumers does not want to be a victim of a e-marketing operation. They are actors with whom we can communicate in a friendly and transparent way.

– Change: technologies and, especially, their uses are changing rapidly. A company must quickly take advantage of new uses to promote its products or services, or risk losing the trust of consum'actors who would quickly switch over to the competitor.

In 2008, Sébastien Soulez[2] had defined e-marketing as a set of operations for Internet marketing, especially to communicate about a product or service in order to promote it. The Internet is now considered a new channel of communication.

This communication channel is made up of a set of vectors [BRI 07] presented in Figure 5.2.

Figure 5.2. *Synopsis of Internet marketing*

---

2 Author of the book "*Le Marketing*" [SOU 08].

– E-advertising (e-pub): in 2007, Joel Moulhade carried out a meta-analysis of several definitions. This enabled him to define e-advertising as follows: "e-advertising technology uses digital information, in particular, all audio-visual and interactive aids to promote Websites services or to inform and convince users to buy a product or service that is either online or offline. This is achieved through the purchase of space and the profitability of Websites".

– Sponsored links: Yahoo Search Marketing defines "sponsored links" as follows: "they are links displayed in a gray box, signifying that they are paid for. Advertisers who have paid through an auction system to get their link are indicated in the shaded area".

– Co-registration: entails proposing two online entries at the same time to a user. Behind the scenes, agreement between a relay site and an advertiser is that when a user signs up for one of its initial offers, a subscription to an offer from the advertiser should be automatically proposed to him – the user is free to accept it or reject it.

– E-mailing: is therefore the mass mailing of the same message via e-mail, with a business objective, to a set of recipients (after segmentation) who are likely to subscribe to the offer [PUB 08].

– Affiliate marketing: this is a vector based on performance marketing. This uses an affiliation platform which becomes a trusted intermediary [GRO 07] between the advertiser and another Website – the affiliate or Website publisher. The advertiser then pays the affiliate in proportion to his contribution to the business. The metric used is CPA (Cost per Action).

– SEO (Search Engine Optimization): the objective of SEO [SAP 07] is twofold: first, register the Website in the search tools database; and second, position the Website among the initial pages of results.

– Viral marketing: often likened to Buzz Marketing [VER 04], this aims to initiate, expand, and maintain a deliberate information exchange flow. This is based on the propagation of information about a brand or product using innovative media supports.

To control the set of promotional activities of different communication channels which form part of the e-marketing field in the context of evolving Web, "Internet Marketers" now have the Web Analytics discipline at their disposal.

Web Analytics help piloting e-Marketing on a site with the measurement and analysis of user behavior, and hence enable the understanding of "consum'actors" (Web 2.0 approach). In addition, they would enable a corporate sponsor to increase the value of its Web investment [WAR 07], and optimize the efficiency of its Website in order to maximize e-business.

5.3. Web Analytics: an essential discipline for an effective e-marketing piloting

The concern for Website audience evolved a few years ago when scripts had emerged for analyzing Web server logs which contained the IP address, browser, and server actions simply via the "Hit" indicator.

In 1995, visit counters were introduced and Web audience software rapidly began to evolve. Some of the business players were specialized in audience measurement and analysis.

However, this approach, though resolutely tool-oriented, does not allow efficient use of Web statistics solutions. In 2002, the annual eMetrics Marketing Optimization conference was organized; the cradle in which the Web Analytics Association (WAA) was born in 2004 [STE 09].

In 2005, Google launched the free Google Analytics service, and democratized Web Analytics.

Since 2005, experts have formalized and developed the concept of Web Analytics. Avinash Kaushik, author of "Web Analytics, an hour a day" and author of the Occam's Razor blog, conceptualizes Web Analytics 2.0 [KAU 07] by employing competitive intelligence which has so far been considered as an indispensable discipline, but now is complementary to e-marketing and Web audience measurement.

5.3.1. Competitive intelligence for Web Analytics

Web Analytics has gradually evolved and has a mutually agreed definition given by the Web Analytics Association (WAA): "Web Analytics involves tracking, collecting, measuring, reporting, and analyzing quantitative data from the Web, with the goal of optimizing Websites and Web marketing activities".

This definition brings back to memory the definition of Watch as provided by Florence Muet in the CERSI[3] report: "formalized and organized implementation of an information system within an organization, for the continuous and dynamic collection, processing and dissemination of information concerning the organization's environment" [BOU 90].

We observe the following similarities between the definition of Web Analytics and that of Watch (basis for competitive intelligence) [BEL 02]:

– enterprise information system is materialized through Web Analytics solutions and tools for campaign dissemination;

– the "collection" phase constitutes "tracking" and "collection of Web data":

    - tracking is done through a pre-planned marking plan and then the insertion of markers is peformed on the desired Web elements. These elements are most often the pages of a Website, links in a newsletter, etc. Some of the Web Analysis solutions still use log files from a Web server. They may be log files of the Apache Web server on a Linux server, etc.,

    - collection entails storing data from the communication vectors, or from the markers placed on Web pages, or server log files. This may involve several hundreds of bytes per second. The organization of the database is important for future data combination which analysts may be required to perform, as part of their analysis;

– analysis is broken down into two stages for Web Analytics: "measurement" and "analysis", which occurs at the end of the process since it is so far non-computerized:

    - measurement includes a restitution format from the key performance indicator. This may be a rate or ratio; the unit can be time, percentage, etc.,

    - analysis is directly related to insight, knowledge of e-marketing strategy of the project, promotional activities, and experience of the Web analyst. This is often associated with data management based on period comparison, and increasingly with a competitive analysis which helps explain some critical variations of the key performance indicators;

---

3 Center for Studies and Research in Information Science.

– dissemination of information is "to report" to the strategic players. There are three types of reports (tactical resources, middle managers, senior officials) [PET 06]:

- report is key because it directly concerns the Web analyst, who must be able to organize the key performance indicators in relation to the objectives and the business model of the site. Based on this, the analyst establishes a precise analysis,

- watch is essentially based on the business environment, while Web Analytics is based on the environment of its Internet operations. Given the technological possibilities and the revolution in practice and trends of Web 2.0, companies are increasingly transposing their activities on the Internet. This rule is especially true for pure-player companies with a business model based on e-commerce,

- continuous and dynamic continual improvement process is involved in all methods of management by Web Analytics [NAE 06].

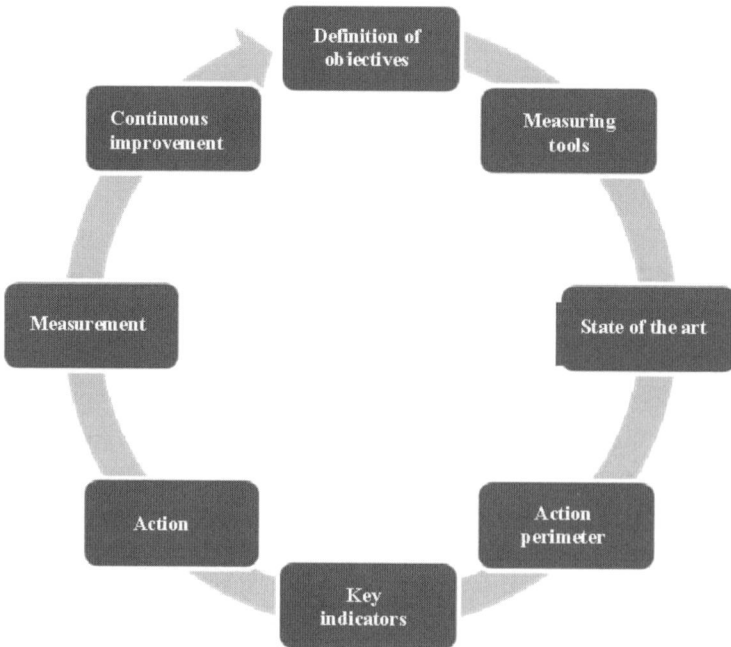

Figure 5.3. *Analyze the performance of your Website in eight steps [NAE 06]*

We can thus infer that Web Analytics include Watch activities for e-marketing decision support. The approach to performance analysis of a Website [LRN 06] presented below shows all the steps with some specificities related to the discipline of Web Analytics. Since the methods are still based on Websites, we will therefore discuss Web Analytics 1.0.

5.3.2. *Web Analytics 2.0: from the approach of Watch adapted to e-marketing to the approach of competitive intelligence adapted to e-marketing*

In 2007, Kaushik Avinash presented a definition of Web Analytics 2.0 at a conference which included several Google experts. The idea of the concept struck him 2 months before the conference, when he attended a series of workshops on marketing ROI.

He has proposed a formal definition which is available on his blog [KAU 07]: Web Analytics 2.0 can be summarized as follows:

– analysis of quantitative and qualitative data from your site and from your Competitor's sites.

– initiation of a continuous continual improvment cycle of the online experience of your customers and potential customers;

– corresponds to the expected results (online and offline).

This new definition demonstrates a new approach based on analysis of data, not only from the Web, but also from the competitors. The concept of expected online and offline results requires a goal at the beginning along with it's new (Web Analytics 1.0). We are seriously concerned about the consum'actor, who is central to the process of continuous improvement.

Web Analytics 2.0 distinguishes itself from its previous version by its amplitude related to new uses of Web 2.0 and indeed the evolution of e-marketing.

The analysis is centered not just on Websites. E-marketer's are rightly focused as much on our own site as the Websites of the competitors. This helps identify the search engine optimization which is not implemented by the competitors and assist in choosing preferred communication channels to generate e-business quickly.

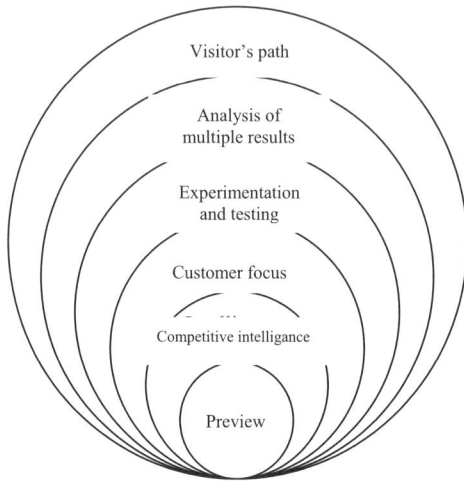

Figure 5.4. *Main components of Web Analytics 2.0 [KAU 07]*

In addition, e-marketing campaigns previously measured at the end of an e-marketing operation via advertising effectiveness indicators are integrated directly into Web Analytics solutions by establishing relationships with Website traffic.

Finally, the reports are dedicated not just to the analyst but to all key enterprise functions for proper dissemination and collective decision-making. Qualitative analyses are preferred over quantitative analyses. Integration of test methodologies (A/B Testing and split testing) works for qualitative analysis.

5.3.3. *Emergence of competitive intelligence tools for e-marketing piloting*

Most of the indicators are from Web services, which are freely available on the Internet by using open interfaces. Some of the vendors of Web Analytics 2.0 solutions have started incorporating indicators of competitive intelligence in their tools, and allow us to explain some variations on a Website by tracking e-marketing activities undertaken by competitors.

5.3.3.1. *Indicators of competitive intelligence for competitive audit of Website and commitment to results*

Although the custom is to never commit ourself to short-term results in the field, it is, however, important for the e-marketing agency responsible for the promotion of an activity, service, or product, to conduct a competitive audit to help quantify the e-marketing strategy which is to be adopted.

This competitive audit can be accomplished through a variety of tools which combine key performance indicators of competitive intelligence.

The International Association of Web Analytics (WAA) had proposed a referential system for Web Analytics in 2008[4], which included most of the key performance indicators essential for understanding and interpreting the results of Web Analytics solutions.

It consists of several categories of indicators, terms of bases, characteristics of visits, visitors' qualification, indicators of commitment, indicators of conversion, and various other indicators.

The current structure shows the integration of many indicators related to converting Internet users into customers, to measuring visitors' engagement and to the qualification of visits, which are the key elements of today's e-marketing. However, we regret the absence of a discussion on key performance indicators of competitive intelligence in a separate section, which forms an integral part of the concept of Web Analytics 2.0 [KAU 07].

We could easily integrate many factors of competitive intelligence online [JAS 08], which would enable understanding of the actions performed, not only on the competitor's Website, but also on our own site [KAU 07], in order to promote the optimization of our e-marketing effectiveness:

– precedence of the domain name: the time when the domain name was created is an important element, since it is taken into account by different search engines. If competing sites have been online for a long time, it will be harder to position ourself against them;

– the position in search engines: there are software programs for position monitoring which gives the position of a site on a search engine relative to a keyword. Since SEO is not instantaneous, tools have been proposed to establish periodic comparisons to follow developments. In this context, it may be interesting to follow up on the positioning of competitors based on keywords which have been included in the meta-keywords, for example.

---

4 http://www.webanalyticsassociation.org/attachments/committees/5/WAA_Web_Analytics_Definitions_20080922_For_Public_Comment.pdf.

This allows us to draw parallels between our evolution development and competitors' development;

– traffic of competitors' sites or that of a leading competitor in an activity: the traffic of a site provides vital information. Some of the tools based on user panels, or on information provided by Internet service providers, offer approximate statistics which can help quantify the traffic generated on competing sites. By considering the trend related to the recent developments and the market, instead of the competition, in order to draw parallels to the variations found on the piloted site, it is possible to audit leading competitors' sites based on the piloted site;

– netlinking quality (incoming and outgoing links): netlinking is essential for search engines, for example, Google uses the Page Rank indicator and Yahoo offers verification tools. Based on this, it is easy to establish an investigation not only for ourselves but also of the competitors. Among the important criteria, we can highlight the quality of the backlink based on the popularity of the transmitting site, the link text (or rather the words in the link text corresponding to strategic keywords), and the context of the link;

– meta-tags: meta-tags contain vital information for referencing a site, but to try this, a strategic review of keywords must be performed, and the competing sites must have passed through this stage. If this is not the case, it is an opportunity to focus e-marketing efforts on organic SEO;

– blog popularity: with the popularity of electronic logbooks, a new discipline derived from Web Analytics has evolved. This discipline has highlighted the excitement of social media analytics; it analyses the social media to generate all the views related to the company, brand, or product [ARS 09]. For competitive intelligence, this could be useful for measuring the influence of communication.

5.3.3.2. *Competitive intelligence tools and solutions for Web Analytics*

As explained by Kaushik in his book, monitoring of the Web ecosystem is essential for understanding variations, identifying the market and, therefore, adapting to our strategy.

There are four types of solutions: solutions based on the user panel (user-centric), site-centric solutions, solutions based on information obtained from Internet service providers, and solutions based on data from search engines.

For performing the analysis, there are hybrid solutions which can aggregate indicators of several types of solutions.

5.3.3.2.1. User-centric solutions

– Alexa can enable us to determine the site traffic based on millions of users. Data is collected through a toolbar installed on the browser. The indicator used is the rate of coverage for 1 million installed toolbars. The interface also offers a comparison service;

– Médiamétrie offers regular publications; this is based on a sample of 7,000 Internet users residing in France.

5.3.3.2.2. Solutions based on site-centric solutions

This category combines tools based on site-centric Web Analytics solutions. Unlike previous tools, these solutions propose measures that are counted and not estimated from a sample of users. Most of the tools are popularized for privacy reasons or to establish a panoramic representative:

– the famous Google Analytics solution, installed on nearly 45.7% of the popular French sites[5] (RESONEO, 2008), offers a benchmarking service that allows us to compare our site's traffic to an average traffic range of site in Google Analytics Benchmark. Similar size segmentation is then performed – the sites are grouped according to the number of visits and divided into three categories: small, medium, and large. When setting up a Google Analytics account, we will be provided with an option to anonymously share our audience traffic. If we agree to do so, then the data are collected to determine the average, giving rise to this benchmarking service. The main indicators for this activity have been proposed: visits, bounce rate, page views, average time spent on the site, pages per visit, and return visits;

– Wysistat solution publisher offers a monthly barometer which consolidates all the audience details on its customers' sites. The barometer is based on more than 1,000 customer sites. It consists of four distinct parts (origin, interest, users, and cross-analysis);

---

5 The study was published on the blog of RESONEO (*société de conseil en stratégie et marketing pour l'e-business*) available at http://www.wagablog.com/2008/12/45-des-top-sites-francais-utilisent-google-analytics/64.

Comparative analysis   30 sept. 2009 - 30 oct. 2009 ⌄

Comparison: Design and development of similar sized websites   ⑦ Open the list of categories   ─•─ Benchmark   ─•─ Your website

2,438 visits
Benchmark: 3,083 (-20.92%)

49.43% Bounce rate
Benchmark: 59.30% (-16.65%)

7,118 viewed pages
Benchmark: 7,713 (-7.71%)

00:02:00 Average time spent on site
Benchmark: 00:01:55 (+4.23%)

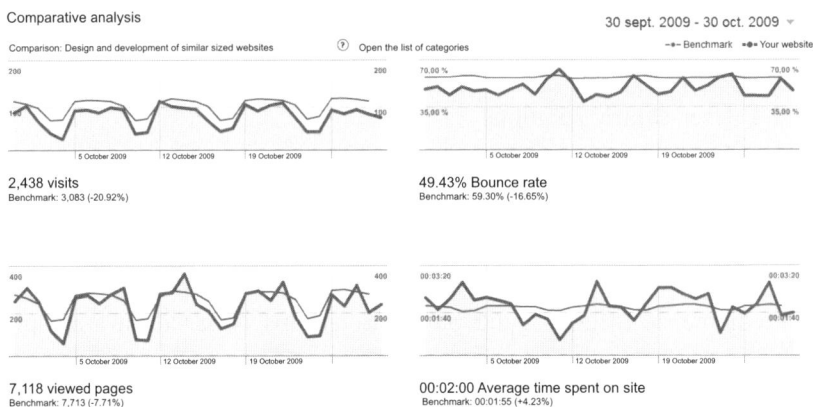

Figure 5.5. *Comparative analysis service integrated into Google Analytics*

– AT Internet Institute barometer (formerly XITI Monitor), proposed by the AT Internet company, has presented publications of various studies on the behavior and facilities of the Internet users, the challenges of Web marketing, search engine barometer and browsers, etc.

– site-centric statistics are based on access log files of a Web server with free access. We can qualify this information as gray information. Indeed, among the solutions of site-centric audience measurement based on the logs, two solutions are distinguished by their popularity. They are Webalyser and Awstats. During installation, the solutions are freely available on the Internet. If the administrator does not specify a particular restrictive action, the statistics can be consulted and accessed freely on the Internet. Since the scripts work in the same way, they can generate the same instructions in their URLs. It is then possible to query Google to bring up the interfaces freely accessible via the command http://www.google.com/search?hl=en&q=site:fr+inurl:usage_200903.html&start=40&sa=N. The command will display open interfaces of the Webalyser solution. It is worth noting that solutions based on server logs are often fault-tolerant to solutions based on markers on the Web pages;

– the number of visits to sites in 10 days using the Weboscope solution from the Weborama solution provider. Weboscope is accredited by the OJD. The sites are classified by theme. Another indicator of popularity was proposed, which is based on votes received by the sites over a period of 90 days;

– audience of sites that request to be certified by OJD. OJD is a French association whose role is to certify the dissemination, distribution, and counting of newspapers, periodicals, and other advertising media such as the Internet. OJD certifies audiences through labeling solutions uniquely based on markers. Only authorized tools can be used to measure the audience of Internet sites seeking certification of their audience.

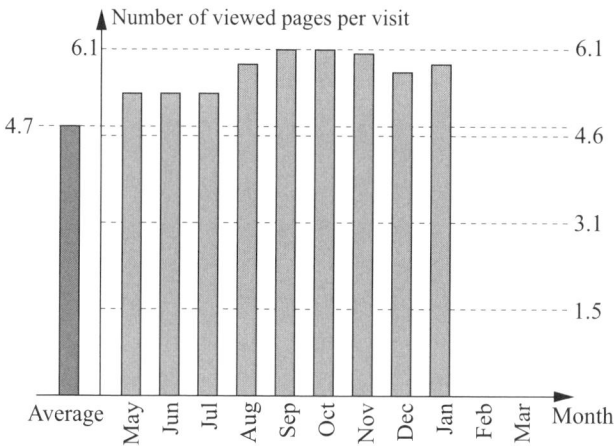

Figure 5.6. *Number of pages viewed per visit from Wysistat barometer*

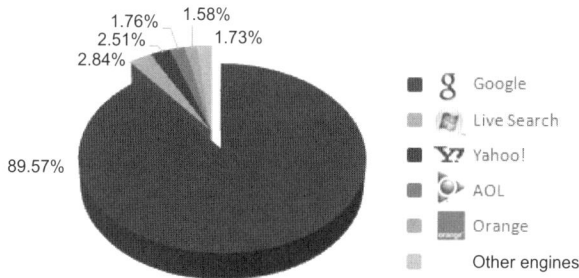

TOP 5 search engines per visit shares
(France - March 2009)*

* Visits generated in France on websites audited by an At Internet solution

AT INTERNET INSTITUTE

Figure 5.7. *Top 5 search engines proposed by AT Internet Institute*

Themes	Visitor classification over 10 days
General	1 **Public iphone**
Current trends	Public iphone mobile telephone
Art and Culture	
Computer related	**Wreport** 109 241 ⇒
Website creation - Computer graphics	2 **Webobo**
Useful information Hints	Create your free website in 5 minutes (Creation of websites - infography)
Languages	
Softwares - Downloads	**Weboscope Free** 68 228 ⇒
OS and hardwares	3 **Clipart-fr**
Sounds - Video	Clipart-fr.com cliparts gif animated (Gifs - Backgrounds)
Mobiles telephones	
GIFS - Background	**Weboscope Free** 61 284 ⇒
Security - Hacking	

Figure 5.8. *Ranking of visitors to sites using Weborama solutions*

5.3.3.2.3. Solutions based on service providers

These solutions are based on analysis of the network traffic of the Internet service providers. With the advent of VoIP and digital convergence, these data become more and more interesting, in understanding behaviors, habits, etc.

To be relevant, the study of various providers is essential. Services allowing access to multiple sources are few and limited. Moreover, they are often focused on the US.

– Hitwise offers studies carried out from multiple providers and representative panels of the Internet users. At present, study reports are available in the United States, United Kingdom, Australia, New Zealand, Hong Kong, Singapore, and Canada.

5.3.3.2.4. Solutions based on search engines

– Google Ad Planner is a solution for establishing media plans and competitive analysis. Although this tool could be classified in the user-centric category, it is a leading search engine that offers this service and offers means of data collection that is so far unknown. Google Ad Planner allows us to make communication plans. It also allows for analyzing the market, identifying the competitors, etc.;

– Microsoft adCenter Labs has offered a series of tools for obtaining details of a Website and its audience. A detector of business intention is proposed to analyze whether the site would be able to generate e-business. A study on the demographic distribution has also been proposed;

– Google Trends for Websites can provide statistics on the frequency with which one or more keywords are typed into the Google search engine. Several graphics are available with demographic segmentation;

– Yooda See U Rank is a position tracking software which enables analyzing and monitoring the referencing of our site and that of the competitors in search engines. SEO Mioche Tool and Rank Checker (Firefox extension) are freely downloadable alternatives;

– Moklic is an interface that enables us to view the advertisers for a given query. This therefore allows for quickly providing a useful overview before driving a campaign of sponsored links. This is used in Google, Microsoft AdCenter, Yahoo Search Marketing, and Miva.

URL: http://www.lab4u.info

Gender: Female-oriented, with the following confidence:

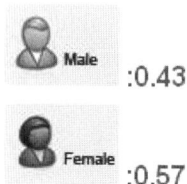

Male :0.43

Female :0.57

Age: 18~24 Oriented with following distribution:

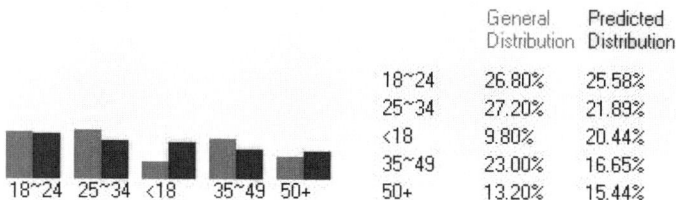

	General Distribution	Predicted Distribution
18~24	26.80%	25.58%
25~34	27.20%	21.89%
<18	9.80%	20.44%
35~49	23.00%	16.65%
50+	13.20%	15.44%

Figure 5.9. *Demographic distribution of Lab4U site proposed by Microsoft ad Center Labs*

5.3.3.2.5. Hybrid solutions

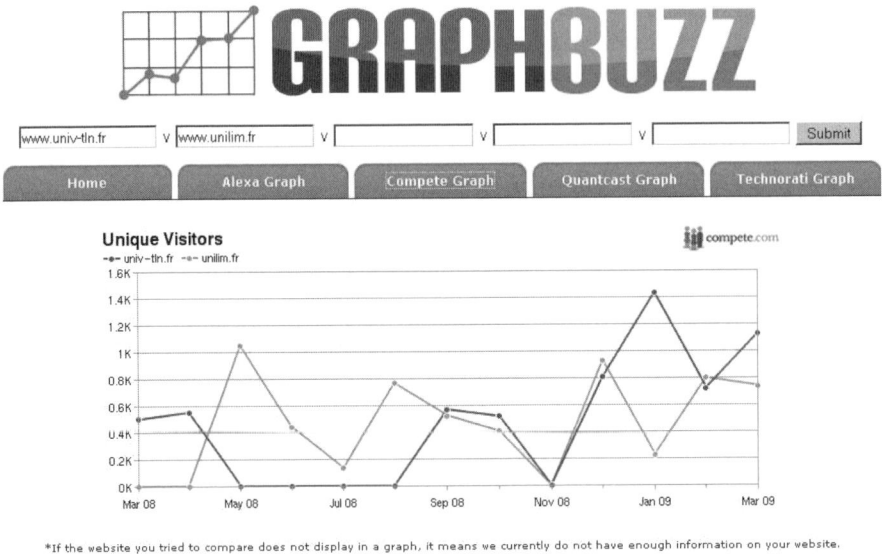

Figure 5.10. *Comparing Websites of University of Toulon and University of Limoges*

– Xinu Returns can test PageRank, Backlinks, indexed pages, positioning, domain name anteriority, Alexa, Compete and Technorati rank, etc.;

– GraphBuzz and Dataopedia are two public interfaces which are capable of aggregating a number of details such as Google PageRank, data from Compete, Alexa, Quantcast, Technorati, etc. The GraphBuzz interface can compare 5 sites of our choice.

5.4. Conclusion

In this chapter, we have discussed the evolution of e-marketing, which is characterized by the Web Analytics discipline for managing and interpreting promotion actions and Web performance content. But the latter (Web Analytics) expands and evolves by incorporating new key concepts to achieve relevance and maximize e-business. Among these new concepts, we note the arrival of competitive intelligence: the true science of intelligence gathering for enterprise competitiveness. This allows for evaluating the strategic positioning of the company and focusing its e-marketing efforts intelligently to deal with competitors' actions and thereby maximize ROI and user experience.

Some of the solutions have already started incorporating this new concept, and independent solutions for competitive intelligence on the Web are emerging. The next step certainly will be the realization of real Web Analytics 2.0 solutions. This can fully incorporate the concept of competitive intelligence, thereby allowing a true "competitive advantage" for an efficient Web control and a strategic interpretation of e-marketing promotional actions.

5.5. Bibliography

[AMO 07] AMOROS S., FRANTZ A., Evaluation des impacts d'un projet en affaires électroniques, HEC presentation, Montréal, 2007.

[ARS 09] ARSON B., Social media analytics: du suivi à la réaction adaptée, blog pages vues et chiffres d'audience, indicateurs et interprétation, 2009.

[AZZ 08] AZZEMOU S., Comment profiter des outils de Web analytique, site web Marketing Direct, 2008.

[BAT 06] BATHELOT B., "Objectifs & Enjeux de la Mesure d'Audience", *Arkantos Consulting*, July 2006.

[BEL 02] BELLON B., "Quelques fondements de l'intelligence économique", *Revue d'économie industrielle*, vol.98, p. 55-74, 2002.

[BOU 90] BOURCIER-DESJARDIN R., MAYÈRE A., MUET F., SALAÜN J.M., Veille technologique: revue de la littérature et étude de terrain, CERSI report, Villeurbanne, 1990.

[BOU 07] BOUTEILLER J., "La mesure d'audience web: une science encore incomplète par Neteco.com", *Neteco.com*, 2007.

[BRI 07] BRIAND A., Le e-marketing en un schéma, Epokhe, 2007 www.my. epokhe.com

[BRI 06] BRITO M., DA SILVA C., "La Règle des 6C du Marketing Révélée", *Arkantos Consulting*, October 2006.

[BRO 08] BRODEUR S., "Introduction à l'analyse Web", *Conférence PHP*, Québec, March 2008.

[BRU 08] BRUYÈRE S., PILLET V., QUONIAM L., "The piloting of e-commerce performance: Development of a model of assistance to piloting by objectives", *Journal of Internet Banking and Commerce*, 2008.

[BUR 07] BURBY J., ATCHISON S., *Actionable Web Analytics: Using Data to Make Smart Business Decisions*, John Wiley & Sons, New Jersey, 2007.

[CAL 03] CALCIU M., "Expérimentation et aide à la décision en Marketing sur Internet", *Pôle de Recherche Marketing, d'EDF R&D*, Association Française de Marketing (AFM), Tunis, May 16, 2003.

[CHE 08] CHEVALLIER N., ROUX C., "Optimiser les performances d'un site web", *Conférence Intellicore*, Valbonne, 2008.

[COU 08] COUSIN H., *Web 2.0 Les mutations d'Internet à l'heure de l'Intelligence Economique*, Association EGE, Paris, 2008.

[CUM 06] CUMBROWSKI C., *Competitive Intelligence and Competitor Analysis of Paid and Organic Search Marketing Activities*, cumbrowski.com, 2006.

[DAS 06] DA SILVA C., "Marketing & Web Marketing: Même Combat", *Arkantos Consulting*, September 2006.

[DAN 08] DANILO R., OLIVIER P., "Web Marketing", *Travail de Séminaire Customer Relationship Management, Chaire d'information Systems*, University of Friburgensis, Switerzerland, 2008.

[DAN 04] DANN S.J., *Strategic Internet Marketing 2.0*, John Wiley & Sons, New Jersey, 2004.

[DEC 06] DECHAMPS R., DE CLERCK J.P., Multi Variable Testing, OX2, September 2006.

[DOU 09] DOURNAUX C., Les fondamentaux du e-marketing, blog, Capitaine commerce 2.1, 2009.

[DUF 98] DUFOUR A., Cybermarketing, professional thesis, CNUCED/OMC, Geneva, 1998.

[DUM 01] DUMAS L., PERREAULT J., PETTIGREW D., *La veille Marketing... Une dimension du système d'information Marketing (SIM) à développer*, ASAC (Association des sciences administratives du Canada), London, Ontario, 2001.

[EQU 06] EQUIPE DE WEB-ANALYTIQUE.COM, Web analytique. L'émergence d'une nouvelle activité, web-analytique.com, September 2006.

[GON 07] GONON I., *Communautés Virtuelles*, University of Limoges, Limoges, 2007.

[GRO 07] GROSS S., *L'affiliation de nouvelle génération*, White Paper, Public-Idées, Levallois Perret, 2007.

[GUI 04] GUICHARD E., BEAUCHAMP S., GARZON M., *Mesures de l'Internet*, Les Canadiens en Europe, Paris, 2004.

[HED 06] HÉDÉ C., "Les 6 Phases du WebMarketing", *Arkantos Consulting*, February 2006.

[HOB 08] HOBEIN E., "Mesurer la performance de la communication sur le web", *Le Journal du Net,* Website 2008.

[HUY 09] HUYGHE F.B., "Web 2.0: Influence, outils et réseaux", *Publications Numériques, Revue Internationale d'Intelligence Economique*, 2009.

[JAN 07] JANSSENS-UMFLAT M., EJZYN A., VANDERCAMMEN M., *Marketing: E-business*, *e-marketing*, *cyber-marketing*, De Boeck, Brusells, 2007.

[JOU 04] JOUËT J., "Les dispositifs de construction de l'internaute par les mesures d'audience", *Le Temps des médias*, no. 3, p. 160-174, 2004.

[KAR 06] KARAYAN R., "Mesure d'audience Internet: comment s'y retrouver?", *Le Journal du Net*, 2006.

[KAU 07] KAUSHIK A., *Web Analytics: An Hour a Day*, Sybex, Indianapolis, 2007.

[LAN 07] LANNOO P., ANKRI C., *E-marketing et e-commerce*, 2nd edition, Vuibert, Paris, 2007.

[LER 04] LE ROUX A., *Les mesures de l'efficacité publicitaire*, e-thèque, Lille, 2004.

[LEN 06] LENDOR C., DA SILVA C., "Promotion Internet: Avantages & Inconvénients", *Arkantos Consulting*, February 2006.

[LEN 08] LENDREVIE J., DE BAYNAST A., EMPRIN C., *Publicitor: La communication 360° on line et off line!*, Dunod, Paris, 2008.

[LOU 08] LOUBAT T., "C'est quoi le webmarketing?", *Indixit Blog Webmarketing*, 2008.

[MAN 08] MANOJ J., Online Competitive Intelligence Factor, Web Analytics World, 2008.

[MAU 06] MAUBLANC H., RENAUD F., *L'e-marketing la stratégie de la performance*, ACSEL/UDA, Paris, 2006.

[MEI 98] MEISTER J., *Corporate Universities; Lessons in Building a World Class Workforce*, McGraw Hill, New York, 1998.

[MEI 04] MEINGAN D., LEBO I., "Maîtriser la veille pour préparer l'intelligence économique", *Knowledge Consult*, 2004.

[MOR 06] MORETTO A., "Défi et définition du cybermarketing", *@toutWebmarketing*, May 2006.

[MOU 07] MOULHADE J., "Les formes de rentabilisations des sites Internet par la e-Publicité", *Cahiers du Lab.RII*, 2007.

[NAE 06] NAEEM A., "Analyser la Performance de Votre Site en 8 Etapes", *Arkantos Consulting*, June 2006.

[NAY 07] NAYLOR D., CUTTS M., SULLIVAN D., "Website analytics vs. competitive intelligence metrics", *SEOGuides*, 2007.

[ORE 05] O'REILLY T., "What is the Web 2.0", *EUTECH The Software Factory*, 2005.

[OZT 06] OZTALAY M., DA SILVA C., "Définition du Marketing Internet", *Arkantos Consulting*, February 2006.

[PEC 08] PECQUET P., "La mesure du trafic entre dans l'ère du Web Analytics 2.0", presented at *Web Analytics, Salon e-Commerce*, Paris, September 25, 2008.

[PET 01] PETERSON E., Web Site Analytics: Key Performance Indicators, Digital Edge Report, September 6, 2001.

[PET 06] PETERSON E., The Big Book of KPIs, Web Analytics Demystified.com, 2006.

[PET 07] PETERSON E., The Web Analytics Business Process, Web Analytics Demystified.com, 2007.

[PET 08] PETERSON E., "Web Analytics: a day a month", Web Analytics Demystified.com, 2008.

[PUB 08] PUBLIC IDÉES, Comment optimiser ses campagnes, emailing b2b? De la préparation à l'analyse des résultats, white paper, 2008.

[RAB 08] RABY G., "Qu'est-ce que le webmarketing, cybermarketing, e-marketing?", *e-marketing et gestion de projet web*, 2008.

[RAI 06] RAIMBAULT T., Les mesures d'efficacité de la communication sur Internet, rapport master recherche, IAE, Lille, 2006.

[RAV 01] RAVOT P., *Le webmarketing*, Hermès, Paris, 2001.

[RIC 08] RICHARD-LANNEYRIE S.C., *Le e-marketing*, Le Génie des Glaciers, Chambéry, 2008.

[ROS 07] ROOS P., *L'évaluation de la performance de la communication média*, CREG, Saint-Ouen l'Aumône, 2007.

[SAL 07] SALVETAT D., LE ROY F., "Coopétition et Intelligence Economique", *Revue française de gestion*, no. 176, p. 147-161, July 2007.

[SAP 07] SAPORTA S., *Référencement sur le net*, Eyrolles, Paris, 2007.

[SBI 09] SBIHI B., "Web 2+: Vers une nouvelle version du Web 2.0", *ISDM*, no. 35, 2009.

[SCH 07] SCHOMANN E., Quels indicateurs utiliser pour l'analyse d'une campagne?, marketing-etudiant.fr, 2007.

[SDL 08] SDL TRIDITION, Marketing Multicannal, white paper, 2008.

[SOU 08] SOULEZ S., *Le Marketing*, Gualino Editeur, Paris, 2008.

[STE 09] STERN J., "The history of the eMetrics marketing optimization summit", *eMatrics Marketing Optimization Summit*, Santa Barbara, 2009.

[TES 07] TESSIER G., Piloter la performance e-business grâce aux solutions de webanalytics – Applications et sites web, Groupe SQLI, December 2007.

[TIS 09] TISSIER M., Internet Marketing 2009, ouvrage collectif, vol. 5, Petit Livre Rouge du Marketing Interactif, EBG, Paris, 2009.

[VER 04] VERNETTE E., FLORES L., "Communiquer avec les leaders d'opinion en marketing: comment et dans quels médias", *Décisions Marketing*, vol. 35, no. 3, p. 23-37, 2004.

[WAI 09] WAISBERG D., KAUSHIK A., "Web Analytics 2.0: Empowering customer centricity", *The original Search Engine Marketing Journal*, 2009.

[WAR 07] WARREN J., "Introduction Web Analytique", presented at *WebCom Montréal: Pannel sur la mesure Web*, Montréal, 2007.

PART TWO

Innovation

Chapter 6

Parallax: Mindset 2.0

6.1. Introduction

This chapter discusses the influence of cultural contingencies in strategic decision-making. It means thinking about science by interlaced reading of psychoanalysis, cultural biology, and professional profile in a systemic and open perspective. In fact, the 2.0 concept does depict a need to take into account the wealth of cultures and individual potential in the digital age. In addition, this wealth is synonymous with creativity and out-of-the-box thinking. Such thinking can express itself fully in strategic decision-making and therefore falls in the gamut of competitive intelligence. We will analyze the multicultural influences made possible by digital technology before drawing out their consequences on competitive intelligence.

6.2. Thought and action in the digital age

The profound changes related to interactive digital era experienced by individual, society, systems, and organizations require professional practitioners in strategic decision-making, and competence and behavioral analysts to cope with the economic intelligence programs. The current architectonic tripod of participation, sustainability, and application of information (shared, collaborative, and distributed) of economic intelligence processes requires professionals with qualities far beyond the knowledge that

Chapter written by Patricia DUPIN.

can be acquired through education or professional practice. The requisite skills fall within a new set of skills and a vision of reality in which we are enclosed – the understanding of the progression of the impact of major social phenomena and also secondary phenomena which reveal a potential (the state of latency) of social applicability. Until the late 20th Century assertiveness of strategic thinking was characterized primarily by vision, analytical accuracy ability, proactivity, creativity, and intuition. The advent of the digital age has added new critical success factors to this set of human capabilities. These factors are:

– systemic thinking: connectivity and analysis of the impacts of the environment and the society on the sustainability of business;

– the ability to see simple: to be able to find simple solutions within complex structures;

– the ability to evaluate the weak signals: ability to detect relevant information in order to build the future from the present;

– the ability to make strategic decisions: to be able to "put ego on the table", i.e. put aside one's personal goal, personal power, and the omnipotence tendency on knowledge. Coming out of one's little world to create a thought capable of amplifying and promoting better solutions, where the ability to share is important.

6.2.1. *Creativity and richness of civilizations*

It is through social transformation – of the community – that new ways of thinking and acting treated here cross the planet, and this, in a more or less intense way, depending on location. This has been evidenced by the results of such massive movement in the great cultural fraternity which represents our planet, and whose bases are located on the four largest and most influential civilizations of our globe: Western civilization, Islamic civilization, Asian civilization, and Indian civilization.

Each of these four civilizations has over 1 billion people living on a multicultural basis. These civilizations represent over 80% of the human population. Each civilization is the collective result of a set of actions based on a different lifestyle. Custom, culture, belief, social economic and political structure, and the use of available technologies show the potential applicability of collective intelligence either at a regional level or at a global level. Collective intelligence relates to the study of mankind that fundamentally structures the society.

Using the concept of Pierre Levy, Collective Intelligence "...is an intelligence that is distributed everywhere, constantly enhanced, coordinated in real time, and leads to an effective mobilization of skills" [LEV 94].

The concept of the anchor point of culture was introduced by Humberto Maturana [MAT 02], a Chilean biologist. Cultures are closed networks of recursive discourse, actions, and emotions: "different cultures imply different mental spaces i.e. different configurations of conscious and unconscious relational/interactional scale of living through different configurations of emotion" [MAR 07]. Cultures are not genetically determined, but their conservation is due to an evolutionary genetic modification. The values involved in these cultures are abstractions of the dynamics of social life, and are more closely related to this emotional domain which constitutes social life. Social reality evolves from the moment it is identified by an observer.

According to Lou Marinoff's analysis, Western civilization is in two columns: the first is the philosophy of Greece and the second, the Judeo-Christian. The intermingling in these cultures existed only in the West with the Romans, French, Spanish, Dutch, Austro-Hungarian, British, and Americans as its main representatives. From this dual foundation flows the love of freedom – a prominent feature of Western civilization, where we find the highest standard of living in the world. Backed by the most modern advances in science and technology, free thinking in civilization has produced major innovations which have revolutionized the world. This freedom prevents that ideas be guided by only one generating principle, as can be observed in other cultures. It promotes competitiveness and is also a source of internal and external conflicts. This kind of emerging collective thinking is characterized by a strong individualism, curiosity, and a certain desire for autonomy.

However, freedom leads to intelligence only if the subject or the society attaches a good meaning to it. Meaningless freedom is a victim of the void. For Castoriadis, the test of freedom was not separated from the test of mortality. A self-reflective organization, which is self-established, would be able to continually question its meanings, and lives in the event of its non-existence. Only from this point, is it possible to create.

For psychoanalysis, the death instinct is the energy that drives human beings to live intensely i.e. even at the risk of dying, and still feel more alive than ever. However, it is the life instinct that protects man from death and leads him toward the search for stability and security.

According to Marinoff, the Indian civilization, which was home to a wide range of spiritual practices, made use of an open mind, and indicated a strong capacity for reconciliation of ideas, irrespective of their degree of divergence. The three Gods recognized by the major schools – Shiva, Rama, Vishnu – represent the creation cycle of a civilization, where one is never in a hurry, since we have all the time in the world in a universe that, according to the belief of the Indians, is going through long periods of evolution. Their ability to assimilate various philosophies seems limitless. They have a vision of a wide/comprehensive world, from where their important contributions in science evolved, like the idea of black holes. The Indian civilization is connected to the Western world, especially in mathematics and science. It is on this spectacular worldview, that their most significant potential relics, and in this context, they can produce the encounter between illusion and what they call "joyful spirit". This type of perception of reality promotes the use of their creative potential, which is particularly characterized by spontaneity, lack of barriers and non-activity. This means that the Indian civilization has a global penetration power which is significantly influential.

Is creativity not fit for every human being? When we observe a self-creation process such as autopoiesis1, we find that creativity is intrinsic to the human race. So what are the forces that can develop creativity in professional fields? How do we foster creativity which is so relevant to the human race? How can we recognize "extraordinary" items in human beings, as compared to its ordinary usage for developing ideas and solutions?

A creative individual has an extraordinary ability which allows him to access dispersive primary mental processes, while connecting to the secondary processes – linear logic. According to Freud [FRE 95], the primary process corresponds to a type of functioning of the psyche that accesses the unconscious (like when we dream) while the secondary process

---

1 The term autopoiesis comes from the Greek auto (self) and poiesis (production, creation). It defines the property of a system to produce itself (and to maintain, to define itself). The term refers to the structural dynamics in unstable equilibrium, that is to say, states organized that remain stable for long periods of time despite matter and energy that pass through. The autopoietic approach of Maturana and Varela [VAR 02] evolved in Santiago, Chile, from the article "Autopoietic Systems" which was presented in a research seminar organized by the University of Santiago in 1972.

concerns the rational and Aristotelian thought type. A creative person retains a greater possibility of access to images, metaphor, accented verbalization, and other forms of expression related to the primary process, in comparison to
an average individual. Inspiration is the manner in which an individual joins the act of abstraction of the primary process with the synthesis of secondary process. So, the scientific genius arises when a simple concrete event inductively associates with a universal class, and generates a conclusive psychic reception. Thus, the conditions necessary to promote the discovery process are non-participation, freedom of thought, ability to dream the future, emergence of critical meaning, availability for exploration, mental alertness, openness, and availability to perceive connections and similarities.

Asian civilization, which is exercised in contemplative thought and the ability to empty the mind, has a sophisticated type of abstract thinking, which is the basis of its strategic capacity. It also masters the Western skills – in science, philosophy, languages, and technology. Given the competitive advantage of the mastery of Western potentialities and the natural use of their strategic capability, China, the Sun of the Asian civilization, seeks to align its decision process and emerge as the next dominant power through conquest of high level assertiveness.

In the last century, decisions undertaken by each of our civilizations had a distinct impact. Today, through the interactivity and connectivity of our century, there are new exchanges and the impact of decisions falls on the globe, beyond the geo-cultural barriers and boundaries, which has brought about a new type of system with a different reaction rate without the possibility of unilateral control. Today, faced with conflicts and convergences, we can testify, in a more or less clear way, to the dissolution of boundaries and cultures, in a total new way. Mankind is creating a global culture through the possibility of more or less free interaction between people of all cultures and beliefs. Certainly, humanity is interconnected with products and services which have outstripped the border barriers. Moreover, this interconnection is mainly due to a breaking consciousness, the center point of man's creative potential. It would be worth recalling the words of Edgar Morin "*we are creating the Gods who, in turn, will create us*" [MOR 05]. Barbarian achievements have built these great civilizations, which stay close to this trend. Achieving success by denying success to another is a malaise that persists in our culture, through an illusionary power.

6.2.2. *Specific aspects, individual potential*

Some of the basic human processes trigger several modes of thinking and acting. The first one lies in the opportunities that we possess according to the manner in which were brought up. We are human beings and, as such, we are limited by some organization and some structure from which we perceive the world. Humans interact with the community where they live based on their needs. History is made and new experiences are lived and therefore, the surroundings and the individuals themselves are changing day after day in a spectacular dance. Contact established with the external world depends on the internal organization and structures which we possess. Hence, through our nervous system we would be able to see the world as we describe it.

According to Maturana and Varela, autopoietic beings are essentially sensitive beings which continually create themselves through their interactions with the environment that surrounds them. Based on what these interactions trigger, individuals create the world in coexistence with each other and with nature. We can say that human beings are rational due to their sensitivity. Rationality is interdependent with sensitivity and the consequences of moments lived in a special and unique way of perceiving the world. In this sense, the more interactions where the individual is accepted as legitimate, the more he would appropriate the world in self-confidence and autonomy. People who believe in the well-being and respect of others and things that surround them have greater openness, which allows them to create new worlds, build strategies, and introduce sustainable solutions.

Man's nature is precisely this possibility in the active, positive and not predetermined sense, to make and to be, other forms of social and individual existence defined by this central specificity, which is creation, by the way that man is created and that he creates himself [CAS 02].

6.2.3. *Contributions from psychoanalysis*

The second process is the desiring essence of all human beings. A human being desires. What is desire? What is the difference between desire and need? How can it render a solution discovery to be of quality? Desire is a lack, a void. It keeps us open; it keeps us moving forward; it "justifies" our actions. Pleasure, a constituting satisfaction (homeostasis) which is increasingly rare in our society, is the moment when an individual has a sense of completeness. This is where the difference lies in humans. Why? Since human demands exceed their needs. It is about the essential and recurring demand of love.

It is a request that cannot be given or received like seeds that are given to birds. A baby, for example, needs to eat just like every living being, but if it is true that he also desires milk, it is his mother's love that he wants most.

Human beings built their identity from each other. They desire each other, and hence the solutions we find are intrinsically linked to other people and therefore directly linked to affectivity. When affective links are weakened, the effective interest in the search for solutions is also weakened, irrespective of the environment where it happens.

Faced with the perception of having little to receive, an individual refuses to act or reduce his motivation for actions. The difficulty in relation to the future is directly related to the fear of being betrayed. It is important to note here that when the social/organizational change is perceived as hostile, threatening, and less emotional, sustainable strategic potential is reduced both in the individual and the community. This is due to the fact that what makes a strategy effective is not really the sureness of its success, but rather the feeling of comfort compared to the risk that it entails, in the sense that it's worth trying, and desire that it works. Man is divided between what he has and what he desires to have, between his biological fragile and perishable structures and the heroic and eternal capacity lived in his imagination. Man can feel like God, imagine his spiritual immortality and produce things that will make him eternal in the history of mankind, thus staying alive after death.

Slavoj Zizek, a Slovenian philosopher and politician, emphasized that the search for objective reality was a false search, and a final ploy to avoid confrontation with reality. This researcher tried to diagnose, beyond appearances, the symptoms of modern sociability, by unveiling its joints in order to update the real hidden desire during a discourse. It is true that in our culture, it is quite traumatic for humans to accept that life is not simply a mundane process of reproduction and pleasure seeking, but that it is lived in service of a truth.

In fact, at present, ideology seems to work in the self-proclaimed post ideological world. Now, we execute our mandates without even accepting them, and also without taking them seriously. There is a barrier: "what makes life worth living is the very essence of life: the awareness of something for which someone would want to risk his own life – you can call it freedom, autonomy, dignity, and honor. It is only when we are ready to take this risk that we are truly alive" [ZIZ 02]. However, "imagination is what allows us to create a world, i.e. it allows us to present something on which, without imagination, we would know nothing and we would have nothing to say. Imagination begins with sensitivity. Imagination and social imaginary are essential characteristics of man". It is in social imaginary that he creates language, institutions, law, and culture. Man is a being who seeks meaning, so he creates it.

6.3. Talent for economic intelligence

6.3.1. *Contributions of the psychological theory types on professional profile*

According to Jung's analytical psychology studies, the behavioral style is derived from the combination of psychological types – where type means a disposition observed in an individual, characterizing him according to his interests, skills, and references.

Carl Jung, the eminent Swiss psychologist, has described these four styles by a combination of preferences – thought or feeling and sensation or intuition. Jung then introduced his concept of introversion and extroversion, thus identifying eight distinct types of behavior. These types were described in the classic book by Jung, "Psychological Types", which was published in 1921. For Jung, "extroversion and introversion are natural attitudes, antagonic to each other, or directed movements, which have already been defined by Goethe as diastole and systole. Their harmonious succession should form the rhythm of life. Achieving this harmonious rhythm would be the highest art of living" [JUN 76].

But no human being is exclusively tagged as introvert or extrovert. The dominant style is characterized by the use of power, a sense of urgency, proactivity and vision potential; influence style, sociable, curious, and spontaneous; the style subject is characterized by the ability to support projects, teamwork, enjoying stability, and the style of compliance, precision oriented, rules and standards, careful and gifted with analytical capacity.

The four styles are applicable to all the individuals, but not all will be obvious. In general, two of these four styles would be stronger. But by the number of aspects obtained from each of the four styles, several types of behavioral schedules would then be possible. The talent of a subject is determined by the establishment of its most important skills, combined with his view of the world and the special interest that motivates him. Based on this, four main styles (pure determinants) have been highlighted, and four others have been derived by combining the first four, therefore leading to combined profiles. Therefore, the eight distinct types of behavior are: conductor, persuader, promoter, relater, supporter, coordinator, analyser, and implementor. In addition to this set of eight pure styles (which can be viewed in the fields), 56 cells appear as possible combinations, and different types of talent can be defined for each, finally totaling 64.

Regarding the dynamics of the profile, the more one gains in a skill, the more he loses in another. This is considered as contrasting competence and combinations between close fields. It leads to persuasive coordinators, directors, and relational styles. When additional skills are required, the design of possible combinations to meet the demand in question becomes more specific.

In the case of intelligence analysts and decision-makers, who should have critical success skill in more than two styles, it is necessary to establish the priorities or the importance of each required competency, so as to be able to draw a viable profile for a human being. It must be noted that a person cannot respond positively to all these qualities. In general, when a person tries to gain recognition in all the areas, he distinguishes himself in nothing and only achieves mediocrity. Obviously, when one of the four fields stands out, the opposite field loses its strength, which can be seen in Figure 6.1.

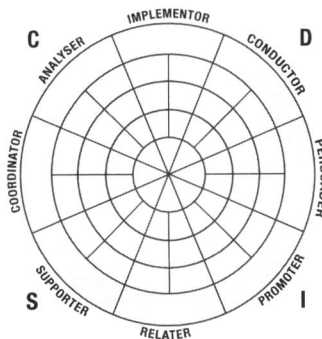

Figure 6.1. *Success Insight* ® *- circle of profiles*

An individual's qualification, training and years of experience demonstrate his knowledge and mastery of some aspects of his professional field. However, this does not allow us to predict his behavior during the process. Knowledge does not define productivity or proactivity, or the ability to establish a good level of intelligence to answer a question or a project, or the ability to make decisions at the right time. Knowledge is not the evidence of intuitive or strategic ability.

In addition, business, market, or cultural characteristics can demand other individual abilities. It is often noticed that a certain profile succeeds very well in one type of organization and fails in another, where the same type of knowledge is required. This occurs since all relational environments have a set of characteristics which, once met, highlight the desire, vision and skills of each, or conversely, neutralize them. This happens more intensely for relief positions. The higher the post, the more mental and less operational the required activities become. The individual must provide specific or global solutions in order to produce an excellent overall performance.

Each profile has its own dynamic feature. This may vary over time, with some degree of flexibility, but it is not recommended that this variation be too large as compared to the natural profile of the person. The farther the dynamic feature is far from the individual's natural profile, the greater would be the effort needed to achieve it. Effort is not synonymous with talent – the spontaneous way of acting of a person reveals a competence that is above the known average.

Here are some of the examples of what occurs frequently. When the analytical capacity is first triggered, the professional would respond less quickly and have a narrower view of the opportunities. However, he would be more specific in the details and would be more meticulous. He may or may not have the capacity for interaction or relationship, according to the second highlighted aspect. If, on the contrary, he is very proactive, he has a sense of urgency and vision – and then, with a low analytical capacity – one would observe an inattentive process, be less cautious, and would be negligent on some of the aspects in case of some complexity, which can be risky for the business. If the ability to influence people comes first, this may indicate an interactive process, information disseminator and persuader. However, because of the low level of strategic assertiveness, we may not see it evolve. Everything depends on the ability that comes out second.

The combination of dominance and compliance with other aspects below average, reveal processes that generally are harder, and in fact do not take into account the interaction and the team. To get a precise combination, we must first be able to find the type of intelligence which is required, and then define the possible ideal profile, by considering the relevant market and the culture in which such a need is present. Currently in the market there are tools to show natural, actual and under pressure profiles with a good measure of assertiveness. Regarding the analysis of teams, the goal may be to establish the skills for collective action or, alternatively, specify the duties and responsibilities which are compatible with each existing profile.

Communication and intuition must now be structural in virtually all the high performance teams. Distributed networks and heterogeneous interactions require agility, flexibility, planning, capacity in decision-making, managing multi-coordinated teams, managing risks and interpersonal relationships. "A social structure based on networks is an open, highly dynamic system, capable of innovations without jeopardizing its own balance. Networks are appropriate instruments for a capitalist economy based on innovation, on decentralization of workers and companies who are devoted to flexibility and adaptability to a social organization aimed toward replacing space and invalidating time" [CAS 01]. Even on networks, new behaviors show a distinct quality: to become smarter over time and react to environmental changes.

In his book "Emergence", Steve Johnson foresees, in a network society, encouraging random encounters and discovery of patterns through signs. He highlights the word "local" as an ideal word for understanding emergent behavior in the logic of networking: think globally, act locally in a collective action which produces a global effect and enhances interaction. "If a picture is worth a thousand words, then an interactive model must be worth millions of words" [JOH 02].

Other studies suggest that professionals working in the information age should know how to transmit or communicate their ideas, in addition to possessing great negotiation skills, understanding of applications and ability to present solutions. "Today, a professional profile devoted to business is ideal. It is essential that the IT professional learn to do teamwork. This includes working with professionals from other fields, which need technological tools to perform their tasks. At this beginning of a new century, dynamism, initiative, and flexibility are desirable attitudes in companies, without ignoring the constant pursuit of learning in line with market needs. ...Success is a result of the execution of tasks to which we devote our best energies, whether intellectual or emotional" [DIA 07].

To identify needs in terms of skills – choose, gather, filter, and process data; provide cross-analysis; select critical information; and generate a record of trends and important things *vis-à-vis* the context – these prospects demand the same set of skills described above, and others such as courage, insight, future vision, balance of thought, and anxiety control. We see next the viability of this coordinated set on a behavioral profile from a performance measuring tool, according to Figure 6.2.

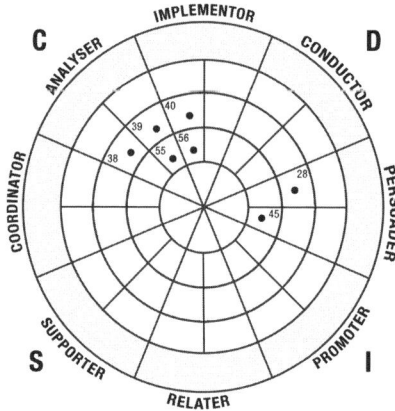

Copyright © 1992,1998. TTI, Ltd., Success Insights Intl, Inc.

Figure 6.2. *Success Insight* ®

In order to understand this picture (represented by the letter C of the Success Insight wheel):

– 38, 39, 40, emphasize more analytical ability and accuracy in an environment, where there is little need to manage people;

– 55, 56, stress on importance on accuracy besides the ability to command and manage small groups;

– 38, 39 and 55 emphasize the primarily analytical field;

– 40 and 56 stress the contractor field (which gives more energy, more strategic capacity and for everybody that is first innovater in this field).

Other boxes located in the C field would demand a deeper analysis of the strengths and weaknesses of the profile, in order to analyze the possibilities of success or failure with respect to the description of the function. For the surface with a focus on dominance (represented by the letter D of the Success Insight wheel):

– 28, presents the importance of the ability to search for results, strategic capacity and innovation. This occupies the field of persuasive profile, with an ability to analyze, and preserve accuracy.

Other boxes located in the D field would demand a deeper analysis of the strengths and weaknesses of the profile, in order to analyze the possibilities of success or failure with respect to the description of the function. For the surface where influence is predominant (represented by the letter "S" in the Success Insights wheel):

– 45, shows the importance of command, negotiation skills, strategic ability, and excellent ability to manage people. This has an ability to analyze and preserve accuracy.

It occupies the persuasive profile field. Other boxes located on the C field will demand a deeper analysis of the strengths and weaknesses of the profile, in order to analyze the possibilities of success or failure with respect to the description of the function.

None of the I (stability) fields are recommended for the task. Adapted profiles (shown as stars in the wheel), which are too far from the natural profile (represented on the wheel as circles) should be avoided.

6.3.2. *Vision and experience of connectivity*

6.3.2.1. *The possibility of being what we see*

Seeing is a common human phenomenon, but the differentiation in what we see is related to how we look at and think of reality and existence – complex ethical explanations are related to thoughts on the reasons for existing in the world. The behavioral profile shows the most predictable behavioral styles, but the way we see and make connections reveal the reactions, motives, and perceptions of reality of each one of us; the belief structure, and values that we are pursuing.

Beck and Cowan [BEC 06], evolutionary science researchers, conducted research on vision. Their results show different levels of evolution being developed like a dynamic spiral, with each level being characterized by a thought style and specific visions depending on each specific need of solution or decision.

These authors have identified eight major systems of values, the first six relate to established paradigms in the field of business, management, education and social communities. The first one includes the values upheld during the process of survival, where the main objective is to stay alive. The second, relying on a circular process, brings a magical and tribal structure, with the sense of enchantment, through which magical solutions are possible. The third, relying on power and exploration processes, has imperial structures whose goal is to win respect, celebrate achievements, and value historical acts. The fourth has a pyramidal structure characterized by authoritarian decision and control process, and whose values reinforce traditional styles, glorifying loyalty. The fifth has a delegating structure, oriented toward autonomy and strategic processes, the search for success, individual recognition, and feeling the need to change to achieve improvements. The sixth has an egalitarian structure, oriented toward consensus, experimental processes, social responsibility, and community. The last two bring new paradigms. The seventh is flexible, integrative, and systemic, and is based on knowledge. The prevalence vision of the world is the wealth of information and multidimensionality. The eighth is holistic. Provided with a global structure and with multidimensional processes and flow, man works in a network, and expands the use of the mind and skills to exist in a globalized reality. Oriented toward embedded systems, he makes effective use of collective intelligence to find solutions for high level problems, without sacrificing individuality.

At present, less than 5% of the population accesses the eighth level of consciousness. However, this type of consciousness functions as the spearhead of collective human evolution.

In accordance with this explanatory proposition, we have a mental levels in a spiral shape in which a level can coexist with another, but each with a given range of limits. Human nature is not static; it changes when these conditions of existence change. Each time a new level of system is activated, the psychology and behavior style change. Life includes all possible levels and then, other levels would evolve. Every individual, firm or corporation can only respond positively according to the level in which it finds itself in the system. By observing the educational, legal and ethical systems, it would be possible to verify the current human existence level, in every civilization, community or individual.

In companies and educational institutions, discourse on systemic and holistic vision is quite common. However, talking about a concept does not imply that one's behavior corresponds to it. Intellectual appropriation usually occurs before the behavioral consistency. In this era, for an intelligence analyst or decision-maker in globalized or transnational companies, it is not sufficient to arrive at the sixth level of the spiral, since the scope of the vision and the specific connectivity at this level do not match the requirements of the Web 2.0. Through individual interviews conducted by specialists, we can determine the level of an individual with a good degree of assertiveness. It is important to analyze the stage where the company lies for which we offer our economic intelligence services, if we wish to understand the type of solution which is expected by the team or the process.

6.3.2.2. *Thoughts of the second order*

Human beings do not see what they did not see, and cannot handle what they have not experienced. Individuals do not see the same things and do not attribute the same meanings to them.

There are also "black" points which prevent us from seeing clearly. We are limited in our organization and structure. So, if an individual does not have a reflecting attitude about life, and if he does not reflect on what he sees, his vision would be restricted to his own finite thinking. The ability of continuous reflection is a desired feature in a professional who claims to be working in the field of strategic decisions of organizations. The second-order thoughts are of primary importance, as regard to the amplitude of gaze. Unlike the first-order thoughts, they do not require prior knowledge or act on objective criteria of truth, and they do not critically depend on the way the circle is created, nor have, as a direct object, something related to the causal chain between the mind and the world.

To access the second-order thoughts, one must reflect on his actions and thoughts through a recursive process by asking oneself questions on issues, in retrospect and by looking "how" to access the perception logic, different from the orthodox logic. The second-order thought enters another field, more expressive than the previous one. Economic intelligence process still requires this skill of reflection.

6.3.3. *Talent management for economic intelligence*

Currently, productivity is derived from innovation, and competitiveness is derived from flexibility. Culture is a source of power, and provides space for the source of capital.

To absorb the benefits of network flexibility, management must foster the dynamics of each element of its internal structure, while seeking to decentralize its units in parallel with a growing autonomy of each.

Human beings are aligned when standards of thought are aligned properly during the meetings. Transparent management facilitates encounters, and maximizes the use of other skills, made free through the well-being. Everything that is hidden or veiled, is felt or perceived, even if we do not know exactly what it is. Any noise in management implies energy usage and dispersion from objective. Confidence is a *sine qua non* factor of success, given its intrinsic connection with the generating motive of human relationships, not to mention the connection with the condition of existence of living beings and autopoiesis.

The product of a strategic project is a "record of importance and trends". This list is not very long, and allows the presentation of relevant and irrelevant issues, which should be seriously considered during the decision process; triggering a demand for a program of intelligence. It is a key product in the management of business risks and success. It is presented in a simple format, despite the complexity it holds.

The methodological form by which we can explain the highest degree of harmony in an intelligence process is the balance of forces. Balancing is the ability of the intelligence plan to possess, in quantity and quality, sufficient relevant information in harmony with the challenge in question, so as to proceed with the analysis through which we can obtain trends and importance. To produce a balance capable of providing a faultless finished product, we must produce a dynamism that can sufficiently balance strengths, in both the internal and external convergence zone and a strategic way to underline the concerned critical points of success and failure. This process is only possible by using collective intelligence: individual talents using the team skills, connected by the art of interaction, through flexibility and amplitude vision, in a learning environment, fostering a high performance for the realization of desires more adherent to human nature, at present and in the future.

Figure 6.3. *Important convergence zones for balancing collective project forces*

6.4. Final considerations

What is the importance of selecting professionals who can see beyond the parameters of mediocrity, break the system and from there observe, act and decide? We should now recall our anchors, our uniqueness, as clarified by Cornelius Castoriadis. This feature of the contemporary world beyond crisis, opposition and profile is insignificant. This characteristic of the masses repeats the same things irrespective of the speech. The immobility of the society is structured in democracy ambush. This insignificance, according to Castoriadis, is complementary to the thought of Hannah Arendt, for whom the contemporary is the perfect decor for the emergence of this historical novelty which we call "being mediocre". Mediocrity does not only concern an individual, it is the hallmark of the masses. This is the new historical novelty that modernity has offered.

For the masses, "none of the superior capabilities of man was necessary to create a relationship between individual life and species life, individual life had become part of the life process, and the only thing needed was to work, i.e. to ensure the continuity of one's life and family. Anything that was not necessary, which was not required by life metabolism in nature, was unnecessary or could be justified only due to peculiarity of human life in opposition to animal life" [ARE 58].

We are concerned here about a singular limiting urgency of the visions of the world and values. This place where life becomes a life process, whose basic need is survival, is a representation of the emergent flux of thought, a support base for bio-psycho-social description. We, thus, have our biological structure, living conditions that surround us, our available intellectual capacity, which drifts in beliefs, motivations, social gatherings, specific objectives: the phenomenon of a system through which we can interpret the world. Something that is closer to conscience than to intelligence – accepting the notion of intelligence as a tool – in an area of congruence with the environment. In interactions related to social life, we change congruently according to circumstances, and we can observe that the congruence of action is in history.

The importance of bringing together professionals having a set of skills which break with mediocrity is to ensure – in addition to technical efficiency – evolutional leaps, enabling great results, competitive differentiation and social consciousness. The leaps are possible when a system of different forces is balanced in perfect coherency with the company. Figure 6.3 shows the difference in trends between ordinary growth and jump. As shown in the figure, it is the balance of forces between a jump and another which leads to the tendency of a certain stability, and which helps prevent failures. In ordinary growth, the company is vulnerable to changes in the internal environment, since it works primarily on internal and external sources, while taking less strategic decisions. Finally, we must consider the fragility of rational explanations. The exercise is to finally accept that no progress is possible without a strong ability of self-criticism.

6.5. Bibliography

[ARE 58] ARENDT H., *The Human Condition*, Calmann-Lévy, Chicago, 1958.

[ARE 93] ARENDT H., *Between Past and Future*, Penguin Books, New York, 1993.

[BEC 06] BECK C., COWAN D., *Spiral Dynamics*, Blackwell, Oxford, 2006.

[BER 67] BERGSON H., *L'évolution Créatrice*, Le Seuil, Paris, 1967.

[CAS 99] CASTORIADIS C., *Figures du Pensable*, Seuil, Paris, 1999.

[CAS 00] CASTELLS M., *The Rise of the Network Society*, Blackwell, Oxford, 2000.

[CAS 01] CASTELLS M., *A Sociedade em Rede; A era da informação: Economia, sociedade e cultura*, vol. 1, Paz e Terra, Rio de Janeiro, 2001.

[CAS 02] CASTORIADIS C., *As Encruzilhadas do Labirinto IV: a ascensão da insignificância*, Paz e Terra, São Paulo, 2002.

[DIA 07] DIAS J., FIAP – Faculdade de Informática e Administração Paulista, São Paulo, 2007.

[FRE 95] FREUD S., *L'Avenir d'une Illusion*, Quadrige, Paris, 1995.

[HUN 97] HUNTINGTON S., *The Clash of Civilizations and the Remaking of World Order*, Simon & Schuster, New York, 1997.

[JOH 02] JOHNSON S., *Emergence: The Connected Lives of Ants, Brains, Cities, and Software*, Touchstone, New York, 2002.

[JUL 05] JULIEN F., *Nourrir sa Vie*, Seuil, Paris, 2005.

[JUN 76] JUNG C., *Psychological Types*, Princeton University Press, New Jersey, 1976.

[LEV 94] LEVY P., *L'Intelligence Collective – pour une anthropologie du cyberspace*, La Découverte, Paris, 1994.

[LIP 06] LIPOVETSKY G., *Le bonheur Paradoxal-essai sur la société d'hyperconsommation*, Gallimard, Paris, 2006.

[MAT 02] MATURANA H., *Transformatión en la Convivencia*, Dólmen, Santiago, 2002.

[MAR 07] MARINOFF L., *O Caminho do Meio*, Editora Record, São Paulo, 2007.

[MAS 02] MASI D., *La Fantasia e la Concretezza*, Sextante, Rio de Janeiro, 2002.

[MOR 05] MORIN E., *Culture et Barbarie Européennes*, Bayard, Paris, 2005.

[SEN 99] SENNETT R., *The Corrosion of Character*, Norton, New York, 1999.

[VAR 02] VARELA F., *Conocer – las ciêncis cognitivas: Tendencies y Perspectives*, Gedisa, Barcelone, 2002.

[WHI 69] WHITMONT E., *The Symbolic Quest*, Princeton University Press, New Jersey, 1969.

[ZIZ 02] ZIZEK S., *Welcome to the Desert of the Real*, Verso, New York, 2002.

Chapter 7

Competitive Intelligence 2.0 Tools

7.1. Introduction

For over 10 years, projects have been carried out in enterprises with the aim of capturing, storing, and sharing knowledge among employees. A range of tools has since emerged. These tools allow us to adapt to a variety of situations such as managing specific projects, codifying expertise and know-how, publishing collaboratively, supporting communities of practice, etc. Facilitating cooperation between individuals separated in time and space and implementing these solutions on enterprise servers can lead to achieving an unusual collaborative potential. The flexibility gained has nurtured the idea of a greater collective efficiency that would result in the emergence of more agile companies which can better read the movements of competitors and anticipate market changes and adapt to them in real time. Collective intelligence would emerge from this potentially permanent interconnection between employees and the knowledge base of their enterprise.

Since, as we shall see later, this expectation was not met by the first generation of tools, it seems to us that services and practices from the "2.0" wave might help companies to achieve this goal. Their characteristics and the modalities of their implementation are sufficiently innovative to the effect that we can speak of a new generation of collaborative tools. We will show that they have the potential of making collaboration in organizations be

Chapter written by Christophe DESCHAMPS.

more natural and therefore, a collective intelligence synonymous with competitiveness could emerge from this simple quality.

We will focus, first, on understanding what Web 2.0 is, why it deviated from the traditional framework of knowledge management in organizations, enabling them to move to a state of collective and continuous watch. Second, we will endeavor to describe the 2.0 tools and services that can be used throughout the stages of the watch cycle. We will be considering the 2.0 technologies from two angles: first, their potential use, which can be collaborative, once deployed on the enterprise server, and second, their exploitation as online services for information watch process. Finally, we will conclude by examining some changes initiated by these technologies, which could impact on competitive intelligence practices in the medium and long term.

7.2. The impact of 2.0 tools on the deployment of competitive intelligence in business

Knowledge management as seen today is dated to over 10 years[1]. Legitimately or not, this has originated primarily from computer science, as confirmed by Aurélie Dudezert. By studying the 24 most commonly cited articles on knowledge management by the Web of Science, she found that more than half were published in journals devoted to Management Information Systems (MIS) [DUD 07].

7.2.1. *The limits of first-generation knowledge management projects*

This technological approach has not yielded the desired results. U.S. research firm Bain & Company has conducted a study every year since 1993 to determine the management tools used by companies [RIG 07]. In 2007, *knowledge management* was listed for the first time among the Top 10 (8th out of 25); however, as specifically identified by the study, the overall satisfaction of users is much lower than that of the nine other tools (22nd out of 25). This reveals, firstly, a very strong expectation of decision-makers toward knowledge management and an increased understanding of the role it can play in organizations, and secondly, a strong frustration with the results obtained.

---

1 Querying Web of Science using the query term knowledge management shows that the citation frequency curve starts from 1997.

In management literature, cautions against the temptation to focus on technology at the expense of users are a popular recurring theme. In fact, many knowledge management projects have not achieved success since they have too often created the "perfect tools" before questioning their future users about their needs. It must also be noted that many software vendors have probably seen, in the publication of Nonaka and Takeuchi [NON 95], a conceptual model for making tacit knowledge explicit, an opportunity to add a strong scientific base to their applications. However, Gold Brynjolfsson and Hitt [BRY 03] have shown that for every dollar invested in the acquisition of information technology to be profitable, companies must spend nine times more to develop their organizational and human capital. The mere deployment of server software, no matter how excellent they are, would be insufficient.

This lack of consideration of actual needs is one of the main reasons for the failure of these projects. In a 2007 study conducted by The Economist Intelligence Unit [CHA 07], 500 executives were asked the following question: "do you think that your business goals are aligned with this statement: give information workers the terminals, software and processes they need to treat multiple sources of data and available information?" The answer was a choice ranging from 1 (perfectly) to 4 (not at all) and 70% of them responded with 3 and 4.

These errors are as a result of non-involvement of employees in the development and deployment of these tools. Michael Idinopoulos, former Chief Knowledge Technology Officer at the McKinsey Consulting firm, explains that "in the old world of e-mails and knowledge management systems, our tools and procedures created a sharp distinction between 'doing one's job' (e.g. write e-mails – NDLA) and 'feeding back the organization (e.g. contributing to the KM management system – NDLA)'" and concludes that "this conception of KM will lead to people spending nearly all their time on e-mails and very little on knowledge sharing..." [IDI 08].

7.2.2. *From Web 2.0 to Enterprise 2.0*

A new generation of technology has, however, impacted on this "old world" and will have strong implications on the way competitive intelligence unfolds in organizations. It is usually described with reference to 2.0 in that it incorporates the principles of Web 2.0.

7.2.2.1. *What is Web 2.0?*

The term was coined by Tim O'Reilly, an organizer of well-respected conferences in the computer world. In September 2005, he wrote an article [ORE 05] titled "Web 2.0", which was presented at a conference held a few weeks later. The article describes what he perceives as the transition from the Web to a new era and he illustrated this transition through examples of online services by comparing the old and the new models. He didn't give any precise definition, but suggested seven principles characterizing the new Web:

– the Web is a platform for delivering services and applications to users rather than a collection of Websites;

– a company must monitor the unique data that are supplied by users of its services and that get richer as they are used;

– users must be considered as co-developers of services;

– collective intelligence must be considered as tools and given the means to emerge;

– the more users of a service, the better the service becomes;

– PC software must be freed by allowing it to slip into objects;

– deployment of flexible and light interfaces based on development methods like AJAX.

7.2.2.2. *Toward Enterprise 2.0*

In an age when companies are concerned about the massive retirement of the *baby boomers* and the consequent risk of losing skills, it does not take long to see them get interested in technologies that can help them better manage, build and share knowledge and expertise of their employees.

To describe this movement Andrew McAfee coined the concept of Enterprise 2.0, which he defines as follows: "Enterprise 2.0 is the use of emerging social software platforms within companies or between companies and their partners or customers" [MCA 06]. He supported this definition by a typology that classifies collaborative technologies into three categories:

– channels: include first generation e-mail and instant messaging that allow everyone to easily create and disseminate information but not to share it on a large scale;

– platforms intranets such as and information portals: unlike the previous category, the content is generated and/or approved by a small number of people but could be read by all;

– knowledge management software (Lotus Workplace, Microsoft Share Point, Documentum, etc.) that has tried to take advantage of the other two types of technology by offering features for capturing and organizing tacit knowledge of employees and providing them with collaborative work environments like groupware.

In a 2005 survey of executives aimed at finding out the technologies of communication and information sharing which are most used in their work, Thomas Davenport noted the predominance of "channels", e-mail in the first place, on the other technologies (of the order of 1 to 5) [DAV 05]. Still, the problems of e-mail in a collaborative environment are real and well known. Besides the fact that it stacks the contributions of all and makes it difficult to search, content is locked in the mailboxes of every employee and cannot be viewed by some of the employees of the company. It is impossible to query them using keywords, so as to discover useful information, at least to know "the people that know" in the organization. Another problem is that the actions carried out on the channels and platforms do not leave any trace in the system and implicit added value that could emerge through aggregation of the user traces does not exist (see below).

McAfee therefore encourages companies to adopt this second generation of technologies that may, for reasons we will now develop, bring them to the next level of collaboration.

7.2.2.3. *More adapted technologies?*

It is worth asking whether Web 2.0 technologies are more responsive to business needs and trying to understand why they could bring collective intelligence where the previous generation has failed. Even though it may seem too early to conduct a comparative study on their effectiveness, it is however possible to propose two hypotheses along the lines of a positive response.

7.2.2.3.1. More adapted to our way of "knowing"

The first, cognitive-wise, was developed by Dave Snowden[2] who has explained that "we know what we know only when we need to know" [SNO 02]. This apparent truism, in fact, expresses the idea that knowledge is always "embedded" i.e. it is specific to one place at a time, but also to the quality of relationships within a group of people.

The context differs for everyone in that it results from the confrontation of an event with the mental patterns of each. For Snowden, knowledge is a way of illuminating a given situation at any given time rather than a higher state of information. Thinking in this sense means that knowledge becomes "actionable" i.e. could be harnessed to support timely decision.

In the same vein, man should no longer be seen as an information processing terminal capable of making rational decisions. The neuroscientist Alex Pouget says that a part of our brain processes in "the background" uncountable data from our environment, recognizes "patterns" (threats, opportunities, etc.) and informs, even without being aware, a probabilistic system of decision-making "that allows us to arrive at a reasonable decision in a reasonable time" [POU 09]. Snowden considers these accumulated bits of information as the basic elements of human knowledge which he describes as "real-time assembly of multiple fragmented memories in a real time context in order to create a new unique application" [SNO 07]. However, as we shall see later, 2.0 technologies are fully responsive to the way we interact with the world. They emit potentially useful bits of information that are in the form of signals, marks, signs that we capture instinctively: one word bigger than another in a tag cloud, a green signal indicating that a contact is online, words more evocative than others in the results page of a search engine, a visual "eye hook", etc.

7.2.2.3.2. Technologies for helping employees individually and collectively

One major difference between these two generations of technology is that the tools making up the second generation were often designed by developers to formulate their own needs[3]. In order to make them known, they have quickly added other features called "social" that made them

---

2 Former director of Cynefin Center for Organizational Complexity, research laboratory on Knowledge Management that depended on IBM. Current CEO of Cognitive Edge company.
3 Joshua Schachter, creator of Delicious, of the iconic Web 2.0 services said: "I originally created Muxway (original name of the service) to manage my own bookmarks".

popular: folksonomies, comments in blogs, rating system, collaborative filtering devices, sharing with "friends", etc.

Hence, tools were originally created to meet individual needs, needs that do not stop at mere personal productivity in the traditional sense, but also include interaction and inter-personal communication. These tools have the potential for, first, solving personal information management problems of: organizing favorites, archiving Web pages or documents, finding them easily, or disseminating them in a broad and targeted way and second, allowing everyone to benefit from the knowledge of the community, for example: when a watcher publishes a post on its company blog, asking for information about a competing product, he actually starts a conversation in the internal "blogosphere" which may eventually contain some answers. Coming back to the distinction made by Idinopoulos, these tools no longer create a difference between "doing one's job" and "feeding back to the organization" since the second type of action becomes a "by-product" of the first and will prove extremely useful to the rest of the company when it comes to identifying internal experts, connecting people working on related subjects or reusing of materials.

Snowden said that "we cannot create a culture of knowledge sharing but we can increase interaction between people and increase their interdependence" [SNO 07]. Brynjolffson showed that employees that are better connected to others were the most productive [BRY 07]. Decentralized, intuitive, both individual and social, blogs, wikis, social networking and micro-blogging applications provide the infrastructure that will allow employees to ask "who knows the answer?" rather than "in which database will I find the answer?".

By emphasizing the connection of individuals to individuals for action purposes (choices, decisions), 2.0 technologies have the potential to generate, as a by-product, the dynamic sharing and "just in time" knowledge to which organizations aspire. In doing so they put in place the foundations for a competitive intelligence which will no longer rely exclusively on a group of individuals trained in a group of CI professionals (and whose functions can then focus on the device steering and competitive analysis) but on all members of the company. The idea is not to chant the slogan "all watchers!" but to allow everyone to "contribute to the conversation" within the company, conversation *a priori* generating enough wealth, explicit or implied, to feed strategic thinking and decision-making.

7.2.3. *Crowdsourcing and RSS: the two 2.0 innovations that make a difference*

The 2.0 tools that we will discuss in the second half of this chapter have a technology in common, RSS, and a principle, crowd sourcing. These two innovations alone justify the idea of a new generation of tools and thus their uses.

7.2.3.1. *RSS: new "grammar" of watch on the Internet and intranets*

RSS stands for Really Simple Syndication or Rich Site Summary. This is an instance of XML that enables easy distribution of content of a Website or blog by "encapsulating" it between the tags (title, author, date, body text, etc.). These will be recognized by software or online services called "aggregator".

The distribution model of RSS information flow is by subscription. Readers just simply need to identify one of the icons below on a site or a blog in order to subscribe to it. From that moment they will not have to visit this site daily as each new article will be automatically distributed in the aggregator.

Figure 7.1. *Sample logos indicating the existence of RSS feeds on a Web page*

Blogs, wikis, social networking services, social bookmarking services, micro-blogging services (see below), all use this technology to broadcast their content. RSS is therefore a true "grammar" that should be mastered by anybody willing to carry out an effective watch on the Web and on different Intranet "spaces".

7.2.3.2. *Crowdsourcing*

This term is a neologism created in 2006 by Jeff Howe and Mark Robinson [HOW 06], editors of Wired magazine. It does not designate a particular feature but a concept applied by 2.0 services for producing an added value by leveraging a large number of users, using a number mechanics channeled through a technical and statistical tool. The produced added value will in turn benefit all[4].

---

4 It also represents one of the conditions imposed by the third and fourth principles of Tim O'Reilly.

This added value can be explicit, when we help Google improve machine translation (Google linguistic tools[5]) or to better understand what images signify by adding keywords (Google Image Labeler[6]), or more often implicit, when we receive suggestions of books that may interest us from Amazon. In this case, the collaborative filtering algorithms[7] of this service use our earlier choices to make automatic recommendations based on the following principle, "if like xx% of our customers, you liked this product, then you might also like this one". The more statistical data the service has about its customers, the more it will be able to refine its suggestions. This "implicit Web" is becoming increasingly important in our ways of creating and "consuming" content on the Web or Intranets.

Alex Iskold, who "invented" this term, explains that: "the implicit Web exists due to the clicks. When we click on something, we are voting. When we spend time on a page, we are voting and when we copy and paste, we vote a little more. Our actions reveal our intentions and reactions" [ISK 07]. Each element of the Web to which we pay attention, gains value due to this simple fact. As explained by Olivier Ertzscheid [ERT 07], we therefore proceed from "an almighty hyperlink, necessarily nodal point for the development of network and related tools and services, to an almighty 'history', of meaningful and oriented navigations".

In a company, crowdsourcing mechanisms allow us, for instance, to alert an employee when several employees with a similar profile distribute or download a document on the intranet. Mechanisms of implicit aggregation (number of hits) can be used to determine the most popular internal blog posts. These mechanisms also allow users to vote explicitly for a reference found on the Web or for an accompanying commentary. Better yet, in a process of competitive watch, they allow going beyond uses based on popularity to a situation of collaborative relevance: classify information, validate it, and analyze it before disseminating it (working with collective intelligence).

---

5 www.google.fr/language_tools?hl=fr.
6 http://images.google.com/imagelabeler/.
7 According to Wikipedia, collaborative filtering includes all methods designed to build recommendation systems using the opinions and evaluations of a group to help individuals http://fr.wikipedia.org/wiki/Filtrage collaboratif (accessed on 01/08/2008).

7.3. Typology of 2.0 technologies for competitive intelligence

It can be seen clearly that 2.0 tools seek to transform the practices of competitive intelligence in an organization. We will consider them from two angles:

– How do we use them for competitive intelligence within an organization (in-house deployment)?

– What uses can they serve as sources of Watch on the Internet?

7.3.1. *Weblogs*

Weblogs or blogs are services for publishing information in the form of notes. They have obvious advantages, as compared to the first generation Websites:

– if one can write an e-mail then one can publish a post on a blog;

– with the comment system, each post becomes a thread to which readers can subscribe and contribute;

– trackbacks enable connecting automatically to follow a discussion taking place on several blogs at once;

– numerous features enable us to find a post (search engine, keywords or tags, archiving by date);

– everyone can follow the blogs of other employees by subscribing to RSS feeds they generate automatically.

7.3.1.1. *Internal uses*

Internal blogs can advantageously replace intranets which are often too static. Deployed at an individual level, they provide some recognition for the ideas, discoveries and reflections of their authors, while allowing the emergence of micro-communities of interest. The watch unit can use them as instruments for disseminating its works (press review, published profiles of competitors, questions, etc.). The conversation that may likely emerge from each post brings added value to the transmitted information.

Besides text, corporate blogs can be used to widely spread audio and video files and can become an interesting medium for this new type of watch (see below).

7.3.1.2. Their uses as information sources for watch on the Internet

In its last study en titled "State of the Blogosphere" [SIF 08], Technorati, a specialized directory, counts more than 133 million blogs on the Internet. 73% of those polled (bloggers) consider blogging as a means of sharing their expertise and experience. These expertises are of course not all of equal value, but it is rare not to find a blog whose author has decided to share his watch results on a sector that correspond to one's own. Watch results can be monitored by subscribing to the blog feed in question.

In addition, some now popular cases[8] have made companies aware of the importance of monitoring about what is said about them in the blogosphere. The ease with which negative information can spread from blog to blog simply by copying and pasting makes it a formidable resonance machine which could harm a company's image in extremely short time.

7.3.2. *Wikis*

Wikipedia, the most famous wiki services, defines it as "a dynamic Website that allows anyone to edit pages at will". It not only allows communicating and disseminating information quickly, it also allows structuring this information, so as to facilitate easy navigation. The word comes from the Hawaiian wiki meaning "quick" or "informal". The advantages of wikis are:

– ease of use;

– many people can work together on the same document simultaneously (synchronous mode);

– it indicates who made a change and at what time;

– prevent the breakdown of data specific to any collaborative work (especially when conducted by e-mail);

– allow the preservation of the project's memory and the various phases of its development through the archiving of all versions of a document (versioning);

– automatically structures content through a particular type of link: wiki links.

---

8 See Kryptonite case in 2004.

7.3.2.1. *Internal uses*

Wikis are designed to allow a team or group to collaborate. For instance at Google, every employee has to work on several micro-projects at the same time in teams comprising three or four people. Each micro-project is based on a dedicated wiki.

A watch unit can use it fully to collaboratively define its watch axes, share collected information and create a new document type, which will be updated as the need arises, while maintaining an historical detailed versions.

7.3.2.2. *Their uses as information sources for watch on the Internet*

It is rare to find a wiki dedicated to a professional theme freely available on the Internet. As such, we cannot really consider them as information sources for watch on the Internet.

7.3.3. *Social networks*

Social networking services or social networks are essential elements of the 2.0 puzzle since they create a link between users by helping to forge and maintain a relationship network. Danah Boyd defined them as "services that enable individuals to (1) construct a public or semi-public profile on a system, (2) articulate a list of other users with whom they share a connection, and (3) to consult and browse the connections of other members of the system" [BOY 07].

They can be considered as the implementation of the famous "six degrees" theory propounded by the sociologist Stanley Milgram in 1967, which postulates that six people connected together are sufficient to make contact with any unknown [MIL 67]. Each user therefore has a detailed profile which, added to others, creates a real directory of experts. A linking mechanism allows each user to know which member of his network is best able to introduce him to a targeted resource person.

7.3.3.1. *Internal uses*

The use of internal social networking features should be considered especially for large businesses. This will help identify the experts of an organization and contact them.

7.3.3.2. Their uses as information sources for watch on the Internet

Networks like Linked In or Viadeo have become first-order information sources for understanding the competition. The profiles that we find can be exploited in various ways (statistical, semantic) in order to draw elements such as implemented technologies (CVs of Engineers), the state of health of the company (number of employees whose status indicates that they are open to job offers), and companies' global organization (services, hierarchies, etc.). Because of fewer restrictions, services such as Facebook are more favorable to the discovery of useful information because employees tend to express themselves on such with less restraint.

Generally, companies should warn their employees about what kind of information should be disclosed – the main issue here is not the likely watch activities that could be carried out by competitors. These services are indeed information treasures for hackers who always are on the lookout for information enabling them to implement social engineering actions[9] in preparation for data theft or dysfunctioning of the computer systems.

7.3.4. *Social bookmarking services*

Gilles Balmisse defines social bookmarking as an action in which "users share their bookmarks and their organization with others. Organization of these bookmarks is usually done through the assignment of keywords, which are called, in this context, tags" [BAL 07]. Tagging has rapidly extended to other services that facilitate online file sharing (photos, videos, office documents, etc.). This new way of ranking is called folksonomy, a term that designates a "collaborative and spontaneous system of classification of Internet content, based on the allocation of key words freely chosen by non-specialist users, which promotes the sharing of resources and can improve information retrieval"[10].

Social bookmarking services are ways to respond collectively to a new need which entails gripping an information mass that becomes more and more difficult to control: the Web.

---

9 Wikipedia defines social engineering as "the act of manipulating people into performing actions or divulging confidential information, rather than by breaking in or using technical cracking techniques".

10 Grand dictionnaire terminologique de la langue française (Government of Quebec's terminological dictionary, www.granddictionnaire.com).

7.3.4.1. *Internal uses*

Deployed on the corporate Intranet, social bookmarking services allow each employee to have a page where he can publish just the interesting pages that he discovers. Just like in blogs, other employees will be able to add comments to his page. In fact a recent development sees enterprise social bookmarking solutions permitting wide publication of both posts and bookmarks/favorites, thereby bringing them closer to blogs. Individual directories so constituted create a dynamic database of watch conducted by each person, which can be exploited in various ways: full-text search, browsing by tags, RSS feeds, filtering by date, document type or author, etc.

7.3.4.2. *Their uses as information sources for watch on the Internet*

Online social bookmarking services like Delicious or Diigo are excellent sources of information. Carefully searching, it is not uncommon to come across people using them to carry out thematic watch and sharing them. Subscribing to their RSS feeds will mean an access to information watch results.

7.3.5. *Micro-blogging services*

These services are designed primarily to address the question "what are you doing now?" The most popular of them, Twitter, allows the user to use 140 characters to convey his/her public or private messages, to those with who the user is in relationship (followers). It entails giving them an idea of what we do daily, evoking moods, interacting, or even relating a conference attended. This feature at first sight seems trivial, but it goes much further than it looks. This feature enables closeness with members of our network and to "humanize" the relationship we have with them. It does not entail daily reading of bits of life, often insignificant, of one another, but basically entails keeping in touch with one another. Over time, the quality of this relationship is transformed into a trust capital, which is indispensable when it comes to actually collaborating on a common project.

7.3.5.1. *Internal uses*

Micro-blogging services are still less prevelant in companies, even though many software servers are already available (see below). It seems, however, that they have an important role to play in the "galaxy" of 2.0 tools. They can indeed be seen as the "cement" of corporate social applications, because they will fill the voids between time-use technologies (blogs, wikis, etc.) and thus ensure continuity in conversation. We believe that they constitute an additional category of tools (i.e. relationship tool) to existing three categories of tools: communication, coordination and production tools [DES 02].

7.3.5.2. *Their uses as information sources for watch on the Internet*

Twitter has become, over a few months, an indispensable source for watch [DES 08]. There are various domain experts[11] who share their findings daily on Twitter. The possibility of entering into a direct relationship through Twitter makes it possible to share information with them and begin discussion as well.

Their ease of use and instantaneity make them effective resonance machines like blogs. The first serious "case" led to the slamming of Motrin Company on the network by a "Twitter community" of angry housewives[12]. Businesses that are concerned about their image or that of their competitors must henceforth take micro-blogging services into account.

7.3.6. *Mashups*

This term can be translated as "mixing", "remixing", or "composite services". It implies using APIs (Application Programming Interface) made available through Web services to combine their content and create a new service. Mashups were popularized through the use of Google Maps API which, as of today, has facilitated the creation of about 1,700 services using cartography.

7.3.6.1. *Internal uses*

In an enterprise, mashups can be created to produce applications that meet a business problem. For instance, we can geo-localize information concerning a competitor and then link them to their profile or recent conversations concerning them in the internal blogs.

---

11 E.g. Kristan Wheaton (Intelligence Analysis), Arik Johnson (Competitive Intelligence), Dave Snowden and Pierre Lévy (Knowledge Management).
12 Details about the Motrin case can be read from the blog fr.readwriteweb.com (December 2008).

7.3.6.2. Their uses as information sources for watch on the Internet

ProgrammableWeb[13], a specialized directory, lists nearly 4,000 mashups available on the Web, which were created from 1,300 APIs. Services that may arise from their combination are only limited *a priori* by the imagination of programmers. For example, there are mashups that automatically put news on a world map according to their origin (Newsmap[14]), there are also mashups that allow "semanticizing" any Web page by identifying names of places or people and enrich them with definitions and pictures (Askjot[15]). There are also those that allow a non-developer to "manipulate" RSS feeds to better filter, merge, disseminate them, etc. (Yahoo Pipes[16]).

7.3.7. *Personalized portals and widgets*

As their name suggests, portals are customized interfaces that each user can configure according to his needs. Once again the model originates from "profane" services such as iGoogle or Netvibes which are aggregators of RSS feeds and can also integrate many third-party services in the form of modules called widgets. There are specifically three types of support for this integration:

– online portals;

– software to be installed;

– blogs: these are pieces of code to copy and paste in an HTML block of a blog.

Irrespective of the chosen media, widgets "embed" and broadcast some of the content of the services from which they originate. The portal thus becomes the interface that centralizes all the information and services to which the user subscribes, in a single view. Thus, there are widgets for a variety of uses: weather, stock quotes, calendar, sticky notes, minicalculator, checking email, task management, etc. In fact every new Web 2.0 service offers its own widgets. In the challenge of not only to better serve our customers, but also to ensure a viral promotion of our brand, good widgets quickly become indispensable.

---

13 www.programmableweb.com.
14 http://muti.co.za/static/newsmap.html.
15 http://semantalyzr.com/.
16 http://pipes.yahoo.com.

7.3.7.1. *Internal uses*

Every employee has at his/her disposal a personalized monitoring board that will enable him/her, for instance, to monitor the news about a customer or competitor, to see the progress of a project, to track the movement of his/her staff on the field, or simply to be notified when a new comment is posted on his/her team's blog. We can assume that each entity of the company will soon offer its own Intranet widgets which the employees can incorporate for monitoring, for collaboration, and for various interaction purposes.

7.3.7.2. *Their uses as information sources for watch on the Internet*

Personalized portals like Netvibes and widgets embedded in them are not sources of information, but are used to disseminate them.

7.3.8. *Summary table of 2.0 technologies for competitive intelligence according to the stages of the watch cycle*

This table shows the use of each 2.0 technology based on the stages of the watch cycle. For the sake of accuracy, we have divided the second phase into two stages: information seeking/retrieval (1) and information capturing (2).

Stages of the watch cycle	2.0 tool	Examples of services	
		Internal	External
Defining watch axis	Wikis	Atlassian Confluence Xwiki	PBworks Netcipia
Seeking and retrieving information (1)	Blogs	Blue Kiwi Knowledge Plaza Jamespot	Search engines: Google Blogs Technorati
Capturing information (2)	Social bookmarking	Blue Kiwi Knowledge Plaza Jamespot Yoolink Pro	Diigo Delicious
	Blogs	Ib.	Ib.
	Wikis	Ib.	Ib.
	Social networks	Blue Kiwi Knowledge Plaza	LinkedIn Viadeo

Analyzing information	Wikis	Ib.	Ib.
	Social bookmarking	Ib.	Ib.
	Blogs	Ib.	Ib.
Disseminating/receiving information	*Social bookmarking*	Ib.	Ib.
	Blogs	Ib.	Ib.
	Wikis	Ib.	Ib.
	Micro-blogging	Yammer Communote	Twitter
	Social networks	Ib.	Ib.
	Portals and *widgets*	Portaneo PersonALL	Netvibes iGoogle
	Mashups	Kapow Intel Mash Maker Mashery SnapLogic	Mashups directory Programmable *Web* Yahoo!Pipes

Table 7.1. *2.0 technologies for competitive intelligence and watch cycle*

7.4. Perspectives of Competitive Intelligence 2.0

7.4.1. *Audio and video watch*

The cumulative phenomena of decline in the price of digital storage, of multiple terminals to create videos and of the development of video broadcasting services (YouTube, Dailymotion, etc.) result in an explosion in the number of audio (including podcasts[17]) and videos files on the Internet. This is a new challenge for corporate watchers who in the nearest future will need to carry out image watch, competitive watch etc. on this file type. It will entail capturing and disseminating them and also storing relevant extracts from them.

Solutions are emerging that allow search engines to index audio files (speech-to-text) in order to do a full text search on them.

---

17 Podcasting is a free means of disseminating audio and video files on the Internet.

7.4.2. *Crowdsourcing – statistical analysis and predictive modeling*

Practices of crowdsourcing generate a constant stream of data that we must be able to handle if we want to extract value from them because, as highlighted by Daniel Kaplan[18], the value "is in the flow, and in that we choose, control, and direct the flow to produce a useable meaning" [KAP 05].

The researcher, Ian Ayres, predicted the advent of what he calls Super Crunchers – individuals who master the statistical analysis of data impacting the real world (stock market, weather, sport betting, etc.) and use them to improve their decisions [AYR 07]. We also observe the massive deployment of online support solutions for data analysis for non-specialists: IBM Manyeyes, Swivel, and Gapminder are among the most popular of them. They allow visualization of statistical data in form of graphs or cloud tags.

Services like Bscopes[19] give a good idea of the kind of information analysis tools we might expect in the years to come. It provides a cartography of RSS feeds that a user subscribes to, and brings out the most popular posts. Each theme in a blog or set of blogs has a visual scalable identity. The user is then able to process more incoming information with the overall vision obtained. These visions will enable him to detect patterns occurring in a part of the map.

Like in weather modeling systems, we may consider modeling this moving universe to be able to predict the "area" from which the next breaking information would be received, which will allow us to discover before others, the innovation that makes a difference with competitive products. This will entail carrying out watch activities by better anticipation.

7.4.3. *Collaborative analysis of information*

If it is obvious that these tools allow discovering new information, the major innovation will, however, be based on the analysis (intelligence analysis). Indeed, if 2.0 technologies facilitate information sharing, they mainly offer mechanisms for treating them collaboratively.

---

18 Director of the Fondation Internet Nouvelle Génération (FING).
19 www.bscopes.com.

Though corporate logic always tends more to rationalization, it is, however, like information, confronted with the subjectivity of those who receive and interpret it. Mechanisms built into 2.0 technologies will allow adding these different subjectivities together in order to co-construct a reality that is always relative but enriched with knowledge of everyone.

Based on this principle, the U.S. intelligence community launched Intellipedia in September 2007, a platform that integrates several 2.0 tools with the goal of helping to consolidate information through validation provided by analysts of various U.S. intelligence agencies.

Companies desiring to implement such devices have the possibility of deploying platforms such as BlueKiwi, or Xwiki Knowledge Plaza (which includes a specific module called Collaborative Watch). We are also beginning to see the emergence of other online services like Rivalmap, fully oriented toward collaborative watch and targeted toward SMEs and open services where everyone can help reconfigure the organization chart of a company (the Official Board) or participate in the elaboration of a SWOT analysis (Wikiswot).

7.5. Conclusion

2.0 technologies have higher potential of adoption than the previous generation of tools because, firstly, they are adapted to the way we interact with the world and secondly, they can be used by individuals without forcing them to collaborate.

This focus on user is accompanied by collaborative mechanisms that "naturally" make them to work toward the benefit of the company. In fact, their actual implementation makes them catalysts of collective intelligence. By offering companies the opportunity to exploit this new field (comprising of the informal and the unstructured), 2.0 technologies allow us to consider new performance levers and new flexibilities that could be seen as opportunities to create a competitive advantage.

7.6. Bibliography

[AYR 07] AYRES I., *Super Crunchers, Why Thinking by Numbers is the New Way to be Smart*, Bantam Books, New York, 2007.

[BAL 07] BALMISSE G., Le social bookmarking au secours de la pertinence de la recherche d'information, author's blog, January 2007.

[BOY 07] BOYD D., ELLISON N., "Social network site: Definition, history and scholarship", *Journal of Computer-Mediated Communication*, vol. 13, no. 1, December 2007.

[BRY 03] BRYNJOLFSSON E., HITT L., "Computivity productivity: Firm-level evidence", *MIT Sloan Working Paper*, no. 4210-01, June 2003.

[BRY 07] BRYNJOLFSSON E., SINAN A., VAN ALSTYNE M., Information, technology and information worker productivity: Task level evidence, Nber Working Paper, no. W13172, June 2007.

[CHA 07] CHADRAN A., "Enterprise knowledge workers: Understanding risks and opportunities", *The Economist Intelligence Unit*, November 2007.

[DAV 05] DAVENPORT T., *Thinking for a Living: How to Get Better Performances and Results from Knowledge Workers*, Harvard Business School Press, Boston, 2005.

[DES 02] DESCHAMPS C., "Les habits neufs du groupware", *Influx*, no. 4, January 2002.

[DES 08] DESCHAMPS C., 26 services pour pratiquer la veille avec Twitter, author's blog, October 2008.

[DUD 07] DUDEZERT A., *Vers le KM 2.0, la recherche internationale en knowledge management*, Ecole Centrale, Paris, March 2007.

[ERT 07] ERTZSCHEID O., Le Web implicite, Affordance blog, July 2007.

[HOW 06] HOWE J., ROBINSON M., "The rise of crowdsourcing", *Wired*, June 2006.

[IDI 08] IDINOPOULOS M., In the flow and above the flow, author's blog, December 2007.

[ISK 07] ISKOLD A., The implicit Web, blog ReadWriteWeb, July 2007.

[KAP 05] KAPLAN D., Le puits et la rivière, site Web Internetactu, January 2005.

[MCA 06] MCAFEE A., Enterprise 2.0: The dawn of emergent collaboration, MIT Sloan, vol. 47, no. 3, April 2006.

[MIL 67] MILGRAM S., "The small world problem", *Psychology Today*, vol. 2, 1967.

[NON 95] NONAKA I., TAKEUCHI H., *The Knowledge Creating Company: How Japanese Companies Creates the Dynamics of Innovation*, Oxford University Press, Oxford, 1995.

[ORE 05] O'REILLY T., What is Web 2.0, design, patterns and business models for the next generation software, author's blog, September 2005.

[POU 09] POUGET A., "Our unconscious brain makes the best decisions possible", Interview, *ScienceDaily*, December 2009.

[RIG 07] RIGBY D., BILODEAU B., *Management Tools and Trends 2007*, Bain & Company, Boston, 2007.

[SIF 08] SIFRY D., "State of the blogosphere", *Technorati*, September 2008.

[SNO 02] SNOWDEN D., "Complex acts of knowing: Paradox and descriptive self-awareness", *Journal of Knowledge Management*, vol. 6, no. 2, May 2002.

[SNO 07] SNOWDEN D., The impact of Web 2.0 on knowledge work and knowledge management, podcast on Wirearchy blog, October 2007.

Chapter 8

Patent Information 2.0, Technology Transfer, and Resource Development

8.1. Introduction

Patents are unique sources of information, since the results published in patents are rarely replicated in other publications. Another peculiarity of patent databases (e.g. World patent database (accessible via the EPO) and American patent database (accessible via the server of the U.S. Patent Office USPO)) is that they are free of cost. This is important to consider in the context of developing countries, even SMIs or SMEs, where financial constraints are increasingly important. Patent information and innovation [AUB 04][1] are almost directly related, since patents highlight existing products and applications in a given domain.

It is also important to note that analytical work (APA Automatic Patent Analysis) cannot be done manually over large volumes. To process and download patents from the "mother" database to a remote computer, we can use an interface (MatheoPatent) which enables performing analytical processing of patents, downloading and updating local patent databases, etc. The subscription fee for this interface is affordable by developing countries,

Chapter written by Henri DOU.
1 AUBERT Jean-Eric, Promoting innovation in developing countries: a conceptual framework, World Bank Institute, July 2004, http://siteresources.worldbank.org/KFDLP/Resources/0-3097AubertPaper%5B1%5D.pdf.

as well as by SMIs and SMEs, and even individuals. It is also important to note that a quality Internet connection is required for this.

8.2. Methodology

Based on a theme related to the natural resources of a country, for example, coconuts (patents are to be queried in English), we downloaded a set of patents ranging from a few hundred patents (family excluded) up to several thousands. Having obtained this corpus, analysis was carried out by using some of the facilities present in the software MatheoPatent[2]. In fact, all the fields present in a documentary record of the patent can be analyzed and combined (networks or matrices), or extracted (thus producing thematic groups) to be analyzed separately.

8.2.1. *Download and analysis*

Figure 8.1 shows the operating process of the interface for uploading and analyzing documents. The analysis is performed by extracting significant elements from all the fields shown in Figure 8.2. These elements are: patent number, date of application, IPC codes (International Patent Classification[3]), country, name(s) of the inventor(s), and name(s) of the applicant(s) (the company that filed the patent, or even the inventor who can also file in his name).

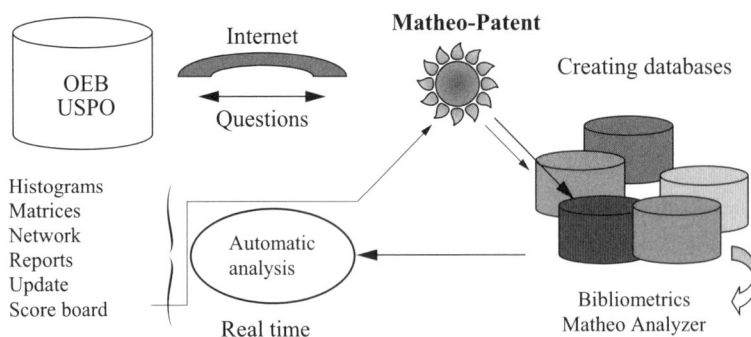

Figure 8.1. *Downloading and analyzing process*

---

2 http://www.matheo-patent.com.

3 http://www.wipo.int/classifications/fulltext/new_ipc/ipcen.html.

IN – Name(s) of the inventor(s)
PN – PR – Patent number, priority number, filling
country (analysis of the first two characters of the
patent number)
IPC – CIB 4 digits or CIB 8 digits
ECLA – Europian classification
Cited Patents
Patent Family

Figure 8.2. *Patent documentary fields*

For more information on the operation of systems for querying, downloading, and automatic management of patents, it is possible to consult the following two references [DOU 07, PAO 03].

8.3. International patent classification

Patents are not given keywords in their bibliographic description as in classical documents. Indeed, keywords would always be increasing and this may not allow doing relevant search due to its diversity and wide scope.

We therefore preferred a different system which is the International Patent Classification.

We divided the field of applications and technologies into a number of classes (letters A to H, which are always at the beginning of a class), then, progressively, by adding letters or numbers to this class, we obtained highly accurate field partitions.

Figures 8.3 and 8.4 give a clear idea of this classification. The classification ranges from one to eight digits, where the digits are accurately coupled.

There are other classifications such as the American classification, the Japanese classification, the European classification, but the one that appears in all the patents is the international classification. This is therefore the one we will use for analysis.

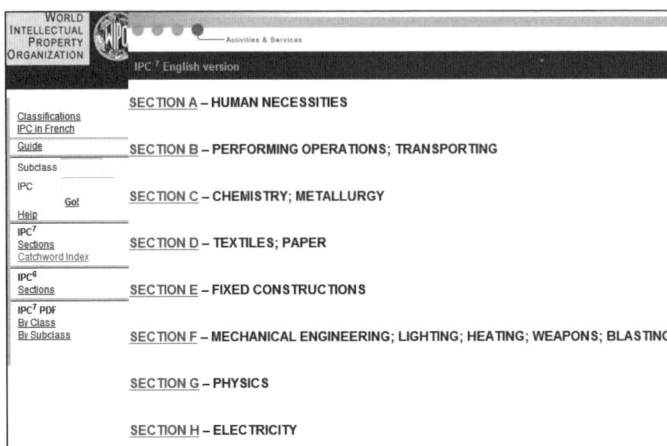

Figure 8.3. *International Patent Classification (extracted from the Website of WIPO, World Intellectual Property Organization) http://www.wipo.int/classifications/fulltext/new_ipc/ipcen.html*

SECTION A – HUMAN NECESSITIES

HEALTH; AMUSEMENT

A 61	MEDICAL OR VETERINARY SCIENCE; HYGIENE
A 61 B	DIAGNOSIS; SURGERY; IDENTIFICATION (analysing biological material G01N, e.g. G01N 33/48; obtaining records using waves other than optical waves, in general G03B 42/00)
A 61 C	DENTISTRY; ORAL OR DENTAL HYGIENE (tooth brushes A46B; preparations for dentistry A61K 6/00)
A 61 D	VETERINARY INSTRUMENTS, IMPLEMENTS, TOOLS, OR METHODS
A 61 F	FILTERS IMPLANTABLE INTO BLOOD VESSELS; PROSTHESES; ORTHOPAEDIC, NURSING OR CONTRACEPTIVE DEVICES; FOMENTATION; TREATMENT OR PROTECTION OF EYES OR EARS; BANDAGES, DRESSINGS OR ABSORBENT PADS; FIRST-AID KITS (dental prosthetics A61C) [6]
A 61 G	TRANSPORT OR ACCOMMODATION FOR PATIENTS; OPERATING TABLES OR CHAIRS; CHAIRS FOR DENTISTRY; FUNEREAL DEVICES (chairs or beds in general A47C; walking aids A61H 3/00)
A 61 H	PHYSICAL THERAPY APPARATUS, e.g. DEVICES FOR LOCATING OR STIMULATING REFLEX POINTS IN THE BODY; ARTIFICIAL RESPIRATION; MASSAGE; BATHING DEVICES FOR SPECIAL THERAPEUTIC OR HYGIENIC PURPOSES OR SPECIFIC PARTS OF THE BODY (methods or devices enabling invalids to operate an apparatus or a device not forming part of the body A61F 4/00; electrotherapy, magnetotherapy, radiation therapy, ultrasound therapy A61N)
A 61 J	CONTAINERS SPECIALLY ADAPTED FOR MEDICAL OR PHARMACEUTICAL PURPOSES; DEVICES OR METHODS SPECIALLY ADAPTED FOR BRINGING PHARMACEUTICAL PRODUCTS INTO PARTICULAR PHYSICAL OR ADMINISTERING FORMS; DEVICES FOR ADMINISTERING FOOD OR MEDICINES ORALLY; BABY COMFORTERS; DEVICES FOR RECEIVING SPITTLE
A 61 K	PREPARATIONS FOR MEDICAL, DENTAL, OR TOILET PURPOSES (bringing into special physical form A61J; chemical aspects of, or use of materials for deodorisation of air, for disinfection or sterilisation, or for bandages, dressings, absorbent pads or surgical articles A61L; compounds per se C01, C07, C08, C12N; soap compositions C11D; micro-organisms per se C12N)

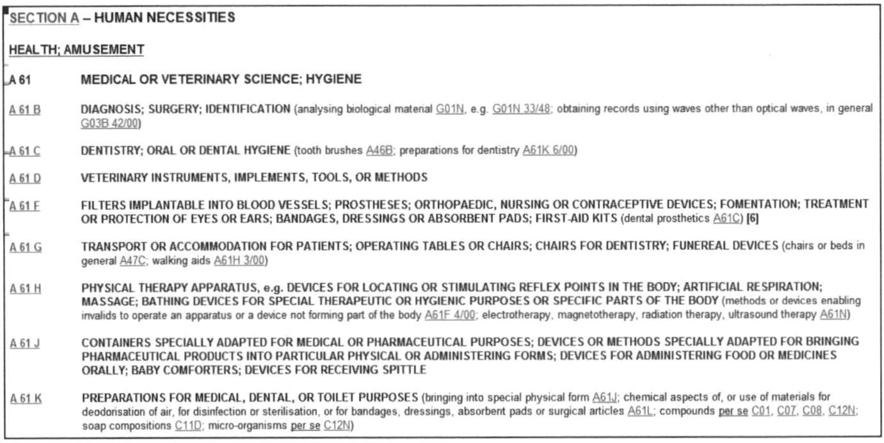

Figure 8.4. *Example of the A61K class extension (here, we have maximum of four digits but more precise information could be obtained by going up to eight digits (see the Website cited in Figure 8.3))*

8.4. A systematic analysis

We will consider an example of a country that produces coconuts. We will perform a classical analysis of downloaded data (thousands of bibliographical references of patents).

Our main information source is the world patent database which can be accessed via the EPO server. Downloading and analysis can be performed via the Internet using the MatheoPatent software.

In the process followed to perform the analysis, we found it interesting to analyze by using a systematic methodology which enables working reproducibly.

This method is useful and is used by various information centers like Toryod in Thailand. We shall refer to this later.

We will proceed through the following stages:

– publication dates: it is obvious that if the number of patents grows over time, it indicates that the subject is an important issue since different companies will pay to protect inventions in this field. This is valid for either the entire patent, or for subgroups that can be created in the downloaded main corpus;

– key technologies used: this is done as above, either on the entire corpus or on created groups;

– main field of applicants: this indicates the major players in the field, and can also be performed on extracted data (according to IPC classes, words in the title or abstract of the patent, dates, etc.);

– key inventors, on the main corpus or on specific groups;

– major filing countries, either at the level of extension of the patent, or at the level of priority country (the filing country of the first patent). A patent may be extended to different countries (it will have different numbers, but it will be the same invention that will be involved). In this case, we will be dealing with a family of patents. Patent families can sometimes be analyzed to determine the policy of the filing company. A sample histogram is shown in Figure 8.5.

This first "vision" of the corpus is certainly useful, but it does not allow making correlations between different previous entities such as knowledge and comparing companies or inventors skills; or performing an automatic benchmarking of companies; or monitoring technologies in time, etc. To achieve such an outcome two methods can be used:

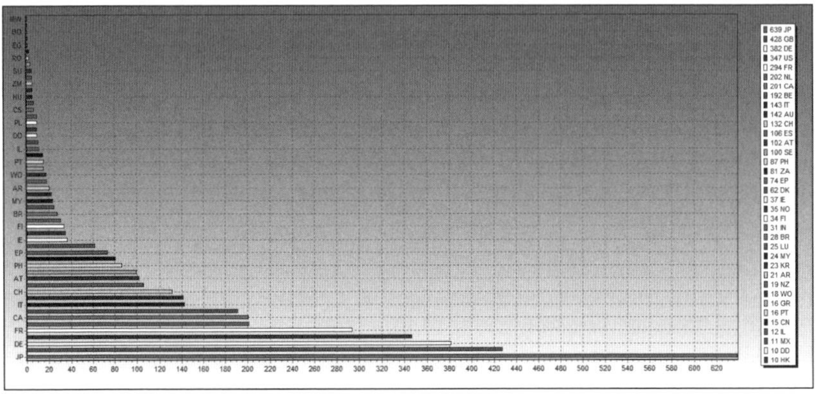

Figure 8.5. *Filing country in the field of coconut*

– the implementation of various matrices between applicants and IPC, inventors and IPC, application dates and IPC, etc. Figure 8.6 shows how to perform automatic benchmarking between various self-filing companies;

– the implementation of networks e.g. network of inventors (i.e. the network of the inventors who appear simultaneously in different patents). Due to the fact that they belong to the same unit, they have links to each other and all these links form a network. This procedure can detect the key technologies in a domain. It can also detect potential key technologies when such technologies appear at low frequencies and do not form part of the main network.

Figure 8.6. *Producing a matrix of correlations between filing companies and IPC to four digits in the field of horticulture*

The analysis of a network reveals the core technologies (those at high frequencies which are nodes of the network), then the low frequency technologies that are either outside the network or as incident not related to the whole network (for example: C03F3/32). These technologies may correspond in some of the cases to innovations.

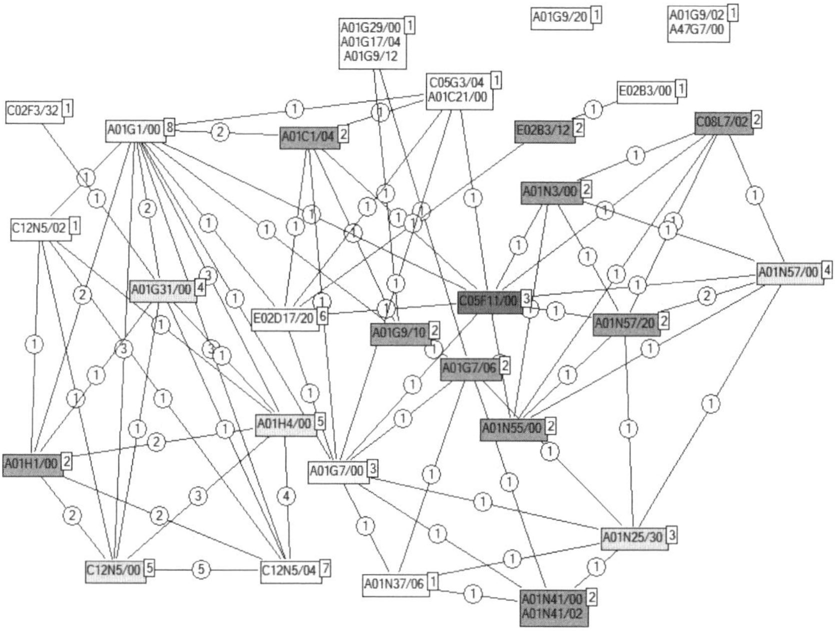

Figure 8.7. *Realization of a technology network in a selected domain. In this case, a horticultural domain (eight-digit IPC). The boxes associated with the IPC include the frequency of IPC in the horticulture group. The frequency of this association is shown in the circles on the branches of the network*

8.5. Search strategies for establishing the initial corpus

It is obvious that if the corpus is not "correct", even a complete analysis would not lead to good results. Therefore, the corpus that will be a good representation of the domain should be carefully established. In this design, we can operate either by performing a broad search and then locally filter (by creating groups) the corpus to be studied, or combine the elements of different fields or use Boolean operators. Figure 8.8 shows the grid for querying by using various options.

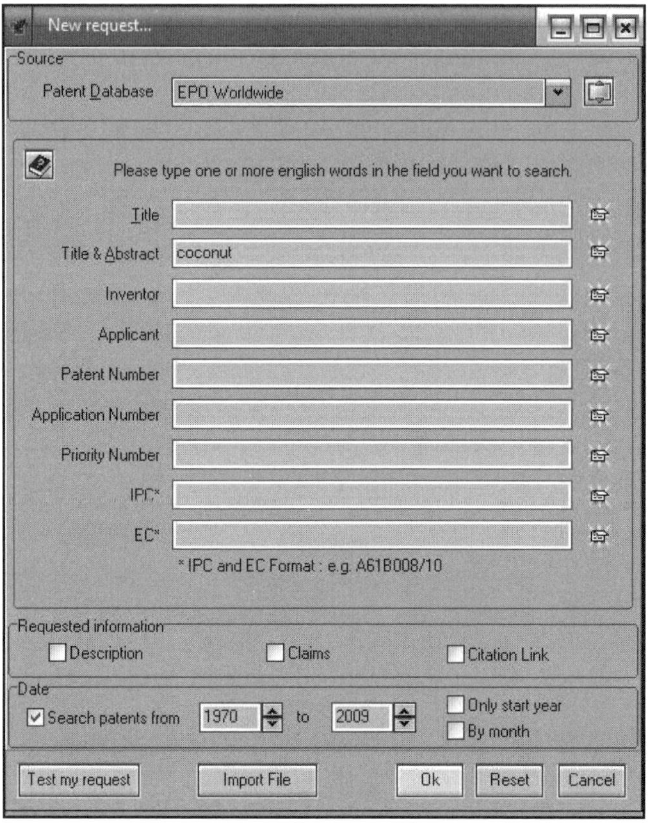

Figure 8.8. *Querying world patent databases*

Here, we queried with the word *coconut* in the title or abstract, over the period from 1970 to the current day. The description, claims, and links to the patents cited by the examiner are not downloaded in order to avoid long download time. Such information is, however, downloaded only for selected patents due to their potential interest for the expert using the Matheo Software. Our data source was the EPO (OPS sister database) Worldwide database (global database accessible via the European Patent Office).

Querying possibilities are the same, whether done directly on the database or by going through the MatheoPatent. However, we recommend that while downloading with MatheoPatent, only bibliographic data (which includes the abstract) should be downloaded and then based on the most significant patents, while other data such as claims or the description of the patent should be downloaded. The software works as follows: we first indicate the number of patents per year, and then we select what we want to download: some years, all or selections per year to test responses.

Once the patents are downloaded, they appear on the computer screen. Clicking on the title of a patent allows us to access the bibliographic description, abstract (if present in the database) and Inpadoc[4] information (if they exist in the database) and the patent family[5].

8.6. Interpretation of results

If the analytical work prepares "the ground" for expert thinking, it is clear that the latter should, according to the results obtained during analysis, guide the choices or recommendations to be made. In this context, and mainly for developing countries and to create added value by developing products, we must find a framework for reflection. This would be generally based on SWOT (strengths, weaknesses, opportunities, and threats) analysis which takes into account the facilities present in the country or region, and at the same time the technological level available to produce certain products or applications. Similarly, general reflection could also be based on Porter's five forces chart, with the central arena representing where companies are fighting, then all around would be forces related to new technologies, new entrants, customer request, and suppliers of raw materials. These various entities can be easily visualized in part in patent analysis, since we have access to technology via the IPC, and then to applicants i.e. potential competitors.

It is also clear that the choices that would be made by experts should be linked to local technological possibilities. For example, if one wishes to develop cosmetic products that some technologies (described in the patent) highlight, for example a supercritical carbon dioxide extraction, it is likely that this cannot be implemented locally. This is important because such analyses, which will be based on reality and not on the wish or the dream of local developers, will enable achieving or at least imagining reliable

---

4 See Inpadoc (International Patent Documentation Center) database. Consulted on May 1, 2009: http://www.epo.org/patents/patent-information/about/families/inpadoc.html.

5 For more information on using these posibilities, visit the following sites: http://ep.espacenet.com/ or http://www.wipo.int/classifications/fulltext/new_ipc/ipcen.html Con sulted on May 1, 2009.

projects. This aspect [DOU 04] is particularly important, since in many cases projects which are from diverse reflections (and this is not only applicable to developing countries alone), are utopian, as these cases are too far from reality.

As the experts' reflections progress, we can complete the corpus. This can be done in two ways: by updating the local database (with the same query) or by adding new references to the local database by using different search strategies. We can also make a combination between the various items presented in Figure 8.2 to update new associations [DOU 05]. For example, we have highlighted simple technologies for developing simple products from coconut, in this analysis.

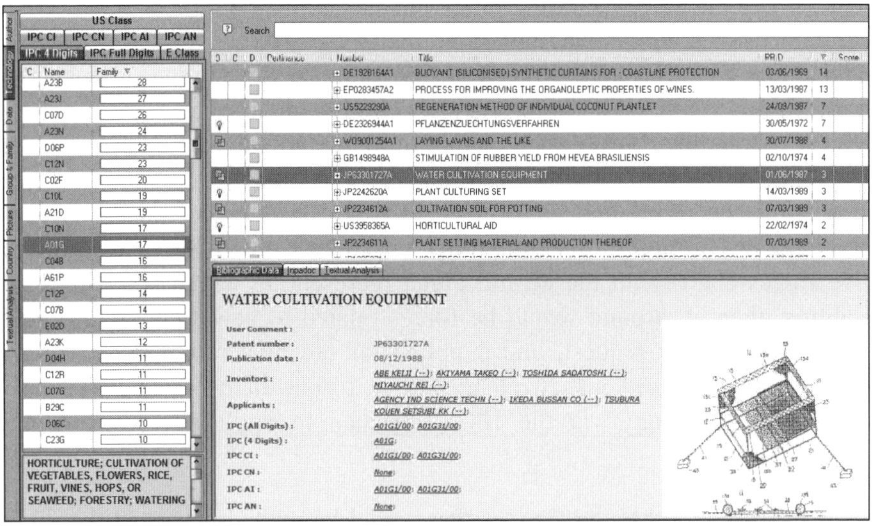

Figure 8.9. *Technological choice by using IPC (in this case, horticulture)*

From these different choices, we can set up groups of technologies by placing in a group all patents related to a product or a selected application. Then we can, for each of these groups, perform analyses either by histograms (applicants, IPC, inventors, dates, etc.) or by making matrices or networks. Figure 8.10 shows different groups of patents selected by the analysis of the patents having the same IPC or words from titles or abstracts in common, as part of the study on coconuts.

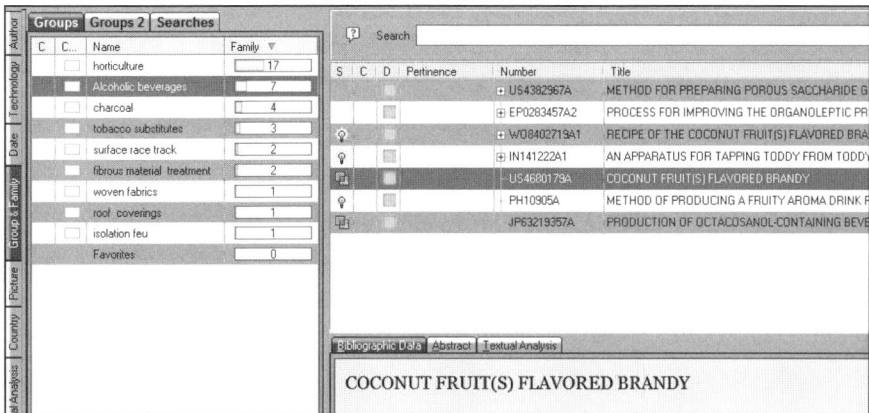

Figure 8.10. *Groups selected for future applications. The choices were made for access to simple implementation technology*

We see that technology choices are made to create products or applications which do not involve very complicated technology. The various fields of the chosen application are:

– selected groups: horticulture, charcoal, tobacco substitutes, processing of fibrous materials, roofing, insulation against fire, etc.;

– another method to realize the selection of patents and placing them in specific groups is to extract significant words from the title and summary. A text parser included in the program allows drawing the histogram of these words and therefore choosing patents that contain them and then placing them in groups;

– one group is selected (horticulture); patents related to this group are presented in the right side of the screen. Choosing a patent provides access to its bibliographic description or its abstract or the words in the title or abstract.

8.7. More precise choices from selected patent

From the selected groups, we can then more precisely select an application like the realization of coir flower pots. The flower pot is lightweight; it can be placed on the ground; it retains moisture and is biodegradable. Figure 8.11 shows the selected application type.

Groups	Groups 2	Searches					

C	C..	Name	Family ▼
■		horticulture	17
■		Alcoholic beverages	7
□		charcoal	4
□		tobacco substitutes	3
□		surface race track	2
□		fibrous material treatment	2
□		woven fabrics	2
□		roof coverings	1
□		isolation feu	1
		Favorites	0

S	C	D	Pertinence	Number	Title	PP.D	▼	Score	Cl.
				⊞ JP83301727A	WATER CULTIVATION EQUIPMENT	01/06/1987	3		
	♀	▦		⊞ JP2242620A	PLANT CULTURING SET	14/03/1989	3		
				⊞ JP2234612A	CULTIVATION SOIL FOR POTTING	07/03/1989	3		
				⊟ US3958365A	HORTICULTURAL AID	22/02/1974	2		
	♀	▦		└ GB1501649A	PLANT CONTAINER AND A METHOD OF MAKING THE SAME	22/02/1974			
				⊞ JP2234611A	PLANT SETTING MATERIAL AND PRODUCTION THEREOF	07/03/1989	2		
	♀	▦		⊞ JP1085071A	HIGH-FREQUENCY INDUCTION OF CALLUS FROM UNRIPE INFLORESCENCE OF COI	24/09/1987	2		
	♀	▦		⊞ FR2556683A1	DEVICE WHICH ALLOWS THE METAMORPHOSIS OF A MAXIMUM NUMBER OF AERI	30/01/1984	2		
	♀	▦		⊞ EP0328451A1	DEVELOPMENT AND CULTURE PROCESS OF A PLANT OBTAINED BY THE GERMIN	10/02/1988	2		
				JP62250999A	HYDROPONIC TYPE PURIFYING FENCE	23/04/1986			
	♀	▦		DE3507429A1	COMPOSITE MAT OF SUCCESSIVE LAYERS OF BIODEGRADABLE, BIOLOGICAL MAT	02/03/1985			

Biblliographic Data | Abstract | Inpadoc | Textual Analysis

HORTICULTURAL AID

User Comment :	
Patent number :	US3958365A
Publication date :	25/05/1976
Inventors :	PROCTOR ATHOL THOMAS (··);
Applicants :	PROCTOR ATHOL THOMAS (··);
IPC (All Digits) :	A01G9/02;
IPC (4 Digits) :	A01G;
IPC CI :	A01G9/02;
IPC CN :	None;
IPC AI :	A01G9/02;
IPC AN :	None;
ECLA (All Digits) :	A01G9/02B;

FIG. 7   FIG. 8

Figure 8.11. *Application choice: hanging coir flower pot*

We note that the chosen patent has also been extended to the UK. This patent dates back to 1976; it is over 20 years old and thus is in the public domain.

In the choices made, one must take into account the protection afforded by the patent. Generally, patents (except for certain patents in pharmacy) are protected for a period of 20 years. Thus, patents that have exceeded this date are necessarily in the public domain.

Patents are also extended to some countries, either through a family or through a global or European patent. In the latter case, the countries where the invention is protected are listed in the patent. If the patent is not extended to a given country or the country is not mentioned in a European or global patent, we can exploit the invention in such a country, but we cannot not extend it (or export it) to protected countries.

For developing countries, the problem is as follows: to create wealth, these countries must export so, if we want to use ideas from patents which have a lifespan of less than 20 years, it is necessary to know if the patent is in use, or if the annuities for maintaining protection are continually paid to patent offices. The latter information is difficult to obtain. Some databases such as Inpadoc, or Litalert (consult the STN list for more information), provide access to the legal status of a patent (though some of these databases are commercial). In some cases the national or European or world office can provide some information, but this will require an information search strategy which is different from a simple document search. A general rule is that, as the annuity increases with time, many patents are forfeited after a few years, except those which are always applied.

There are other examples of choices made from patents that are more than 20 years old. The basic idea of using coconut fibers is to provide protection against soil erosion or to facilitate the growth of plants or grass falls within the common domain.

Examples over 20 years after the patent date, with the described idea and the product, are now in the public domain. Only the product produced could eventually be protected. Hence, from this basic idea of protection against erosion, different products have been developed, as shown in Figure 8.12.

Figure 8.12. *Safety net against soil erosion developed by a company in the Philippines*[6]

8.8. Generalization of the method

This method of working, if applicable in developing countries, can also be extended to the SMEs and SMIs, and even clusters. To illustrate this generalization, we will highlight how an information center in Thailand uses the method and the software described in this chapter to provide innovation assistance and strategic information to Thai companies.

---

6 Information obtained from http://www.alibaba.com/product-free/100611423/Coir_Geotextile/showimage.html. For more information on this type of product which can be exported from Philippines, consult this link: http://*www.alibaba.com/product/roycefood-12299886-0*/productdetail.html.

Figure 8.13 shows the representation of the various studies carried out in implementing the previously described method. We would see its application in different fields: coconut, nanotechnology, water jet massage, palm oil, etc.

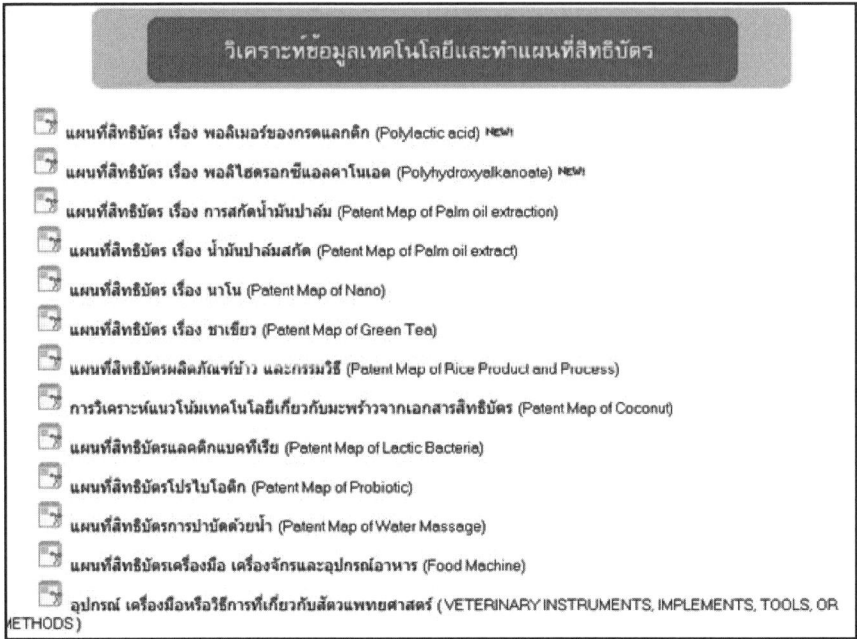

Figure 8.13. *Generalization of the method by a Thai strategic information center*[7]

In order to properly fix the ideas, we will present some excerpts from one of the analyses (patent map or mapping) conducted as part of a study on coconut.

We will detail the source of information, the software and the method (excerpt from the presentation on the mapping of patents on coconut).

ได้ทำการสืบค้นข้อมูลสิทธิบัตรนานาชาติ   จากสำนักสิทธิบัตรยุโรป   ที่ http://ep.espacenet.com   และสำนักสิทธิบัตรอเมริกา   ที่   http://www.uspto.gov ด้วยระบบการสืบค้นและทำแผนที่เทคโนโลยี   ด้วยโปรแกรม   *Matheo-Patent* (http://www.matheo-software.com) (as it appears in the text).

---

7 Available on the Internet via http://www.toryod.com/publicationmapping.php.

A classical mapping operation is presented in Figures 8.14 and 8.15.

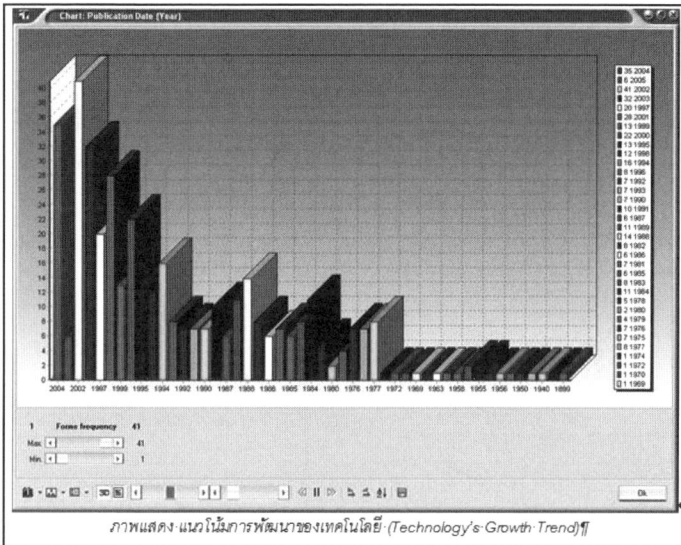

Figure 8.14. *Chronology of the number of filed patents (strong growth)*

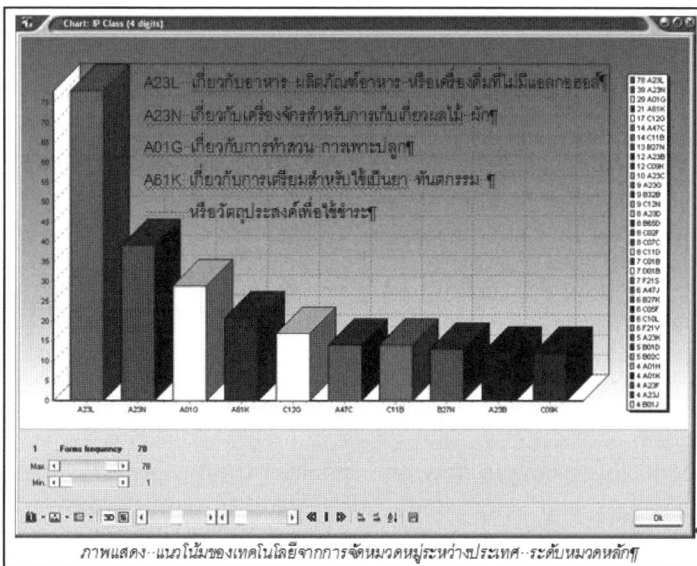

Figure 8.15. *Main classes (for the queried area of application) chosen by the Thai Information Center*

This strong growth reflects that the "coconut" topic is evolving and involves a growing number of patents and companies. In Figure 8.15, we have highlighted the main areas of application and development with the histogram of IPC (in this case in four digits).

Figure 8.16 shows an excerpt of a patent which may have useful applications for the Thai industry and where relatively simple technologies are used. We will see, in turn, a machine to remove the fiber from the coconut, and then a realization of garden pots with the same fiber. The two designs shown are extracts from two different patents.

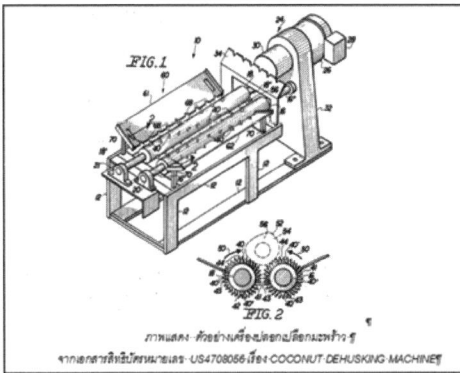

Production of fiber from coconut

Using these fibers in horticulture

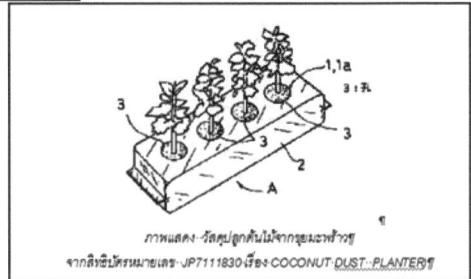

Figure 8.16. *Two examples dealing with the production of coconut fibers and their application to make horticulture pots*

We see in this example how one can start from a general theme (in our case "coconuts") and select application areas (usually based on IPC and taking into account the technology levels available in the company or region), and combining them together to achieve a product that can be exported to the international market. The pots used in horticulture are lightweight, inexpensive, and do not demand much energy for their production, and are also biodegradable while retaining moisture.

We can see that this is a sample application which is in line with the consumer perception: sustainable development.

8.9. Conclusion

In these examples, we wanted to show a method, starting from some of the applications, which would generate a reflection, as well as ideas for new products or new applications. The use of patent information is interesting since it contains both the drawings and language which can be understood by business and R&D researchers. This is an important advantage. In addition, it uses validated data since they have resulted in patents. Thus, they are concrete applications and products which are protected. The fact is that in a patent, we have data on companies, on technological areas, and on dates (and timelines) which enable us to map the strengths and the weaknesses of companies and also detect opportunities and threats (of companies that are already in the sector). Moreover, beyond the simple SWOT analysis, the APA provides food for thought to experts, and also provides the basis for developing concrete projects. The use of existing technologies or transfer of these technologies in different domains will save valuable time for developers. In addition, an overview of what is protected in a given sector may suggest new ideas and thus be a vehicle for innovation [KAB 94, OEC 02].

The only constraint is to know if the patent is not more than 20 years old or if it is still in force or not[8] i.e. if the instalments to be paid by the applicants to maintain protection are always paid or not, or if trials were held regarding the validity of a patent and what was the resulting judgement[9].

The application domain of this method is broad, and concerns coconut as well as various natural products such as wicker, cloves, some natural fuel, various natural oils, biodiesel, etc. We have experienced this way of working in different countries with different natural resources and in all the cases, this

---

8 For more information visit http://www.patentlens.net/daisy/patentlens/261.html.

9 This is the case for the non-free database Litalert accessible via the Dialog server (File 670) which relates only to judgments in the United States and is related to the U.S. Patent Office.

has enabled moving from dream to reality and building reliable projects, the first step in a rational development.

8.10. Bibliography

[AUB 04] AUBERT J.E., Promoting innovation in developing countries: A conceptual framework, World Bank Institute, July 2004, http://siteresources.worldbank.org/KFDLP/Resources/0-3097AubertPaper%5B1%5D.pdf.

[DOU 04] DOU H., MANULLANG D.S., "Competitive intelligence and regional development within the framework of Indonesian provincial autonomy", *Education for Information*, no. 22, June 2004.

[DOU 05] DOU H., LEVEILLE V., MANULLANG D.S., "Patent analysis for competitive technical intelligence and innovative thinking", *Data Science Journal*, vol. 4, p. 209-236, 2005.

[DOU 07] DOU H., DOU J.M. JR, "Bibliometry technique and software for patent intelligence mining", *Managing Strategic Intelligence. Techniques and Technologies*, Mark XU (ed.), IGI Global, Hershey, 2007.

[KAB 94] KABILA I., GUELLEC D., "Le brevet: un instrument d'appropriation des innovations technologiques", *Economie et Statistique*, vol. 275-276, p. 83-94, 1994.

[OEC 02] OECD 02, *Dynamizing National Innovation System*, OECD, Paris, 2000, publié en 2002, http://www.cipi.gob.mx/Biblioteca_Digital_CIPI/Bibliografia Basica/innovacion/Dynamising%20Nacional%20Innovation%20Systems.pdf.

[PAO 03] PAOLI C., DOU H., DOU J.M. JR, MANNINA B., "La constitution d'indicateurs brevets par domaines technologiques", *Bulletin de Documentation Belge*, no. 2, 2003.

Chapter 9

Industrial Property: Competitive Weapon 2.0 (Case Study of Tenofovir)

9.1. Introduction

The Brazilian Law no. 9313 of 13 November, 1996 [BRA 96b], guarantees free and universal access to HIV treatment. Since 1997, Brazil has embarked on an extensive program of free and universal access to antiretroviral drugs (ARVs). The aim of this program is to try to establish a balance between respect for industrial property rights and social rights:

– industrial property rights in Brazil are particularly driven by the Industrial Property Law no. 9279 of 14 May, 1996 [BRA 96a]. This law was reformulated to comply with the minimum standards for the protection of intellectual property, which was laid down by the World Trade Organization (WTO) [BEL 01], [EST 01];

– respect for social rights guaranteed free and universal access to assistance.

The government has managed to significantly reduce the average annual cost of antiretroviral therapy from $6,240 per patient in 1997 to $1,336 in 2004. To achieve this, the government mobilized the state-owned laboratory

Chapter written by Wanise Barroso and Joachim Queyras.

Farmanguinhos[1] and other public laboratories[2] to launch the production of copies of ARVs which were not patented in Brazil before 1996, the ratification date of the Industrial Property Law no. 9279 and the country's alignment with the agreement of the WTO, the Agreement on Trade-Related Aspects of Intellectual Property Rights (TRIPS) [OMC 05a].

Today, the combination of two factors could result in a substantial increase in health spending for maintenance of the program of universal free access to ARVs, i.e.:

– the Brazilian state treats a growing number of patients (nearly 170,000 in 2005 against 20,000 in 1997) [CHE 05];

– there is a change in the international legal environment, particularly through the development of trade relations as laid down by the WTO [OMC 05b].

The creation of a multidisciplinary network of Health, Pharmacy, Chemistry, Intellectual Property, Information and Communication Sciences professionals, as well as policy-makers and non-governmental organizations could solve the problems resulting from the lack of information among the various communities which are affected by the problems of public health policy in the fight against AIDS in Brazil. In this context, it is very important to know about the drugs and their Active Pharmaceutical Ingredients (API) and their patents, records of drugs in specialized agencies (FDA[3], ANVISA[4]), scientific and technical publications. On the other hand, it is necessary to determine the appropriate systems for the recovery, storage, and information management. With this complete system, i.e. experts and tools, it is possible to reduce the cost of the policy on access to assistance for patients suffering from HIV and fill up the gap resulting from the delay in terms of its ability to synthesize the API [MOA 03]. This strategy fits into existing structures in Brazil in terms of information and communication on AIDS, including the *Department of STD, AIDS and Viral Hepatitis*[5] in its

---

1 Institute of Pharmaceutical Technology/Farmanguinhos. Available at http://www.far. fiocruz.br/
2 Pharmaceutical Laboratory from Pernambuco State (LAFEPE), Popular Medicine Foundation (FURP), Ezequiel Dias Foundation (FUNED), Vital Brasil Institute (IVB), Pharmaceutical Laboratory of Alagoas (LIFAL), Pharmaceutical Marine Laboratory (LFM) and Pharmaceutical Chemistry Aeronautical Laboratory (LQFA).
3 Food and Drug Administration, Available at http://www.fda.gov/.
4 National Agency for Sanitary Surveillance, Available at http://www.anvisa.gov.br/.
5 Department of STD, AIDS and Viral Hepatitis, Available at http://www.aids.gov.br/.

various components in close association with the Ministry of Health in Brazil [MIN 05]. The objective of this chapter is to show how an effective policy of monitoring and managing information about drugs and their API, including patent information in terms of content as well as procedure, can help governments, particularly in developing countries, to establish a sonsistent public health policy from the point of economical and geopolitical situation in the globalization era [EUR 06].

9.2. Current status of the subject in the international context

Developing countries and members of the World Trade Organization, were required to bring their legislation in line with the agreements on Trade-Related Aspects of Intellectual Property Rights (TRIPS), adopted in 1994 under the Uruguay cycle of GATT, by 1 January, 2000[6] [OMC 05c]. These provisions require all the signatory countries to protect pharmaceutical products and their manufacturing processes by patents.

The implementation of these agreements, that restrict competition, would prohibit the production of "second line" generic drugs, which were recently patented. However, these drugs are necessary due to the increased number of patients who have developed resistance to "front line" antiretroviral (ARV) or accept the latest recommendations, particularly of the World Health Organization[7] for the treatment of AIDS disease [COR 03]. In addition, the Brazilian state is already spending about 80% of its budget in the access universal program and free ARVs, to purchase imported drugs (62.5% of the $310 million budgeted for the acquisition of ARV in 2005 was devoted to the acquisition of Efavirenz, Tenofovir, and Lopinavir only). The implementation of these agreements has also resulted in restricting the free circulation of the API that are used for the ARV drugs, and so on, the local production of these drugs is seriously threatened.

To solve this problem, Brazil had intended to enforce the Declaration of the President of the General Council of WTO [OMC 03], on 30 August, 2003, which had implemented the Doha Declaration on TRIPS and public health adopted on 14 November, 2001 [OMC 01a, b]. The declaration has

---

6 World Trade Organization, TRIPS. Available at http://www.wto.org/french/tratop_f/trips_f/trips_f.htm.
7 WHO and HIV. Available at http://www.who.int/hiv/en/.

established that "the TRIPS Agreement does not and should not prevent the members from taking measures to protect public health and should be interpreted and implemented in a way that supports the right of WTO members to protect public health and, in particular, to promote access to drugs for all" [ABB 02] [ANR 04].

The TRIPS Agreement provides, in effect, compulsory licensing (CL), especially when a government allows a third party to produce the patented product or using the patented process without the approval of the assignee. But initially, the company needs to try negotiating a voluntary license with the patent assignee on reasonable commercial conditions before requesting a compulsory licensing. If the efforts with the voluntary license fail, the compulsory license would be required, in this case, the assignee must receive the royalties.

The TRIPS also provides for the parallel imports possibility i.e. a non-counterfeit product imported from another country without the permission of the intellectual property owner. However, it must ensure that low cost drugs provided for poor countries are not being diverted to other markets, since this would compromise the differential pricing [HAM 02].

However, there is no legal framework for the implementation of these "flexibilities" [OMC 05d]. This is not a case of legal and economic difficulties relating to the application of CL [ANR 05]. There is, however, under these conditions, the possibility to propose compulsory licensing while negotiating with the intellectual property owner.

Brazil has achieved significant price reductions from the ARV patent assignees, e.g. the reduction was 40% for Nelfinavir and 59% for Efavirenz[8]. However, the WTO's decision of 30 August, 2003 on the implementation of section 6 of the Doha Declaration on the TRIPS Agreement and Public Health[9] was intended to facilitate access to medicine in the developing countries [ANR 05]. This amendment "will allow any member country to export pharmaceutical products made under compulsory license for that purpose. Such countries could change their own laws".

---

8 Health Ministry of Brazil. Available at http://www.saude.gov.br/.

9 Implementation of section 6 of the Doha Declaration on the TRIPS Agreement and public health. Available at http://www.wto.org/english/tratop_e/trips_e/implem_para6_e.htm.

9.3. Research and results on Tenofovir

9.3.1. *Generalities*

As a preliminary, we studied the Tenofovir – Viread®[10] drug, among the drugs prescribed in the treatment and prevention of HIV in Brazil, and so, we decided to present an opposition to the patent application[11] (PI9811045-4[12]) in *National Institute of Industrial Property* - INPI in Brazil, according to the article 31[13] and the article 8[14] (requirements of novelty, inventive activity, and industrial application) of Law no. 9.279/96. The active ingredients's drug is tenofovir disoproxil fumarate (TDF) and belongs to the group of nucleotide analogs. It is prescribed for patients who have strong side effects when an antiretroviral cocktail is being administered.

The Brazilian Ministry of Health had included Tenofovir in the recommended treatment for HIV, from the second semester of 2003. This is the 15th drug belonging to the retroviral class whose use is promoted in Brazil. In 2004, the Health Ministry bought 1,989,510 tablets for a price of US $7.68 each, representing a total cost of US $15.28 million in 2004. This cost was reevaluated to US $25 million in 2005[15]; the only measure that has been taken so far to contain costs associated with Tenofovir therapy was to ask the patient to perform a genotyping test to ensure he/she has no resistance to the drug.

Tenofovir (Viread®) is marketed by Gilead and was approved by the FDA on 26 October, 2001 and by the ANVISA on 26 September, 2002. This drug is sold in Brazil by United Medical Ltd.

A detailed methodology was developed for the study of Tenofovir patents. We have broken this study down into points of equal importance,

---

10 Information available at http://www.viread.com/.

11 Patent cancellation application (*opposition*) is available at http://www.intelliflux.info/ANRS/PROJET/subsidio1.shtml.

12 Patent application is available on the INPL patent database. Available at http://www.inpi.gov.br/.

13 From the time of filing a patent application to the end of its reviewing, interested parties can present documents and information to assist in the verification (*Opposition*).

14 Article 8 of the Law no. 9.279/96 states that "invention which satisfies the prerequisites of novelty, inventiveness, and industrial applicability has the potential to be patented".

15 Information provided by Brazilian Ministry Health.

and this will be the basis in developing the methodology that we present in this chapter.

9.3.2. *Discussion of tenofovir analogs*

The term "Tenofovir" is generally used to name the active ingredients of Viread®, but it is important to view the term in this context, since it appears in the following three active ingredients of different chemical formulas:

– Tenofovir, also known as PMPA;

– Bis (POC) tenofovir, also known as bis (POC) PMPA or tenofovir disoproxil;

– Tenofovir disoproxil fumarate, also known as bis (POC) PMPA fumarate or Tenofovir DF or BPPF or under the trademark Viread®.

9.3.3. *Object of the invention on the patent application*

Patent Application PI9811045-4 refers to the compound (fumarate bis (POC) PMPA or BPPF), which is administered orally to patients infected with the virus or at risk of viral infection. The invention seeks to protect the composition of BPPF, synthesized from the association between bis (POC) PMPA and fumaric acid in 1:1 proportion.

The patent application also seeks to protect other creative elements such as the method to obtain BPPF and the method of oral administration of an effective amount of the composition to a patient infected with the virus or at risk of viral infection.

9.3.4. *The state of the art*

This part of the study is devoted to the analysis of bibliographic references related to the search for previous works, which are composed of scientific papers, patents and data on chemicals which can be utilized in this study. The information was obtained by using different databases[16] and

---

16 SciFinder Scholar (http://www.cas.org/SCIFINDER/SCHOLAR/); Medline/PubMed Database (http://www.ncbi.nlm.nih.gov/entrez/query.fcgi?DB=pubmed); MeS HDatabase (http://www.ncbi.nlm.nih.gov/entrez/query.fcgi?db=mesh); PubChem Database (http://

different Websites[17]. This chapter will describe only the conclusions of the analysis of each article or patents, and not a detailed description.

9.3.4.1. *Chemical analysis*

This analysis would permit us to deduce that:

– there are three substances (active ingredients) under discussion in this work: Tenofovir (PMPA), tenofovir disoproxil (bis (POC) PMPA) and tenofovir disoproxil fumarate (tenofovir DF, bis (POC) PMPA fumarate, or Viread ®);

– it has been known since 1993 that enantiomer (R)-PMPA has an inhibitory effect on cell transformation induced by HIV replication greater than the (S)-PMPA, and that derivatives (R)-PMPDAP[18], (R) PMPA, and (S)-FPMPA[19] are potential antiretroviral agents *in vitro* and *in vivo* [BAL 93];

– PMPA pre- and post-treatment of simian immunodeficiency virus (SIV) in macaques is efficient and (R)-PMPA [BIS 96] is more effective and less toxic than PMEA[20] and AZT[21] [CHE 95];

– (R)-PMPA is effective on infected macaques for at least 19 weeks when treated at 30 or 75 mg/kg, administered subcutaneously in single daily dose for four weeks, and the reappearance of the virus in case of interruption of treatment has been noticed;

– bis (POC) PMPA was selected as a pre-medicine having very good oral bioavailability in monkeys;

– bis (POC) PMPA has a notorious anti-AIDS activity, particularly through its action on human lymphocytes, with less toxicity than the PMPA [BIS 97];

---

pubchem.ncbi.nlm.nih.gov/); Web of Science (http://scientific.thomson.com/products/wos/); Derwent (http://www.derwent.com/).

17 *US Food & Drug Administration*, FDA (http://www.fda.gov/); *European Patent Office*, ESPACENET (http://www.espacenet.com/); *National Institute of Industrial Property*, INPI (http://www.inpi.gov.br/); *Department of STD, AIDS and Viral Hepatitis (http://www.aids.gov.br)*, ANVISA (http://www.anvisa.gov.br/); *United States Patent Trademark Office*, USPTO (http://www.uspto.gov/); ChemFinder (http://chemfinder.cambridgesoft.com/).

18 (R)-9-(2-phosfonilmetoxypropyl)-2,6-diaminopurine.

19 S)-9-[(3-Fluoro-2-(phosfonilmetoxy)propyl]adenine.

20 (9-[2-(Phosphonomethoxy)ethyl]adenine).

21 Azidothymidine (zidovudine).

– the bioavailability of bis (POC) PMPA is much better than that of the PMPA [FRI 97], [NAE 97].

9.3.4.2. *Analysis of the precedence*

This study is the consequence of the analysis of patents after the search for prior art on patent application No PI9811045-4. This study helped us to observe that:

– PMPA has been known from the technical point of view since 28 February, 1989 (Patent No. US4.808.716);

– PMPA and bis (POC) PMPA are protected in the US through patent no US4.808.716 and US5.922.695, respectively, but these compounds (active ingredients) were not protected in Brazil and therefore belong to the public domain in Brazil, and it is possible to manufacture these two substances in this country;

– from a technical point of view, obtaining PMPA as a salt by using organic acid (Patent No. WO9403467) is a technique that was already known since 1963 (Patent No. US4.258.062) and it is impossible to invoke the inventiveness of obtaining a (POC) PMPA in salt form, such as fumarate salt, as specified in the patent no PI9811045-4;

– increasing the bioavailability of drugs by obtaining their respective physiologically acceptable salts, especially the fumarate salt, is known (Patent No. GB942.152, DE2.111.071, US3.682.930, DE2.305.092, US3.994.974, BE859.425, US4.258.062, EP79.545, US4.430.343, EP240.228, US4.879.288, and EP164.865 US5.155.268), therefore, obtaining fumarate bis (POC) PMPA does not involve any inventive activity from a technical point of view;

– Philip L. Gould has written in his article "Salt selection for basic drugs" (International Journal of Pharmaceutics, 33, 201-217, 1986), that the active ingredient in a free base can be converted into pharmaceutically acceptable salt without any physico-chemical and biological modification;

– by searching the databases, it was possible to identify seven drugs[22] registered by the FDA, which are the drugs based on fumarate salt, apart from bis (POC) PMPA discussed in this work.

---

22 Clemastine Fumarate (Sabdoz Ltd.), Ketotifen Fumarate (Bourquin et col.), Formoterol Fumarate (Yanouchi Pharmaceutical Co. Ltd.), Bisoprolol Fumarate (Merk Patent Gmbh),

Analysis of this information has confirmed the results obtained by the analysis of patents and scientific references. This means that:

– the compound fumarate bis (POC) PMPA does not involve any inventive step since the antiviral activity of tenofovir and bis (POC) MPA was already known (Patent No. US4.808.716);

– techniques used to improve the bioavailability of bis (POC) PMPA by causing a reaction between the active ingredient and an organic acid to obtain a salt of this compound, were already known before the filing of the patent, and this technical knowledge is very well-known to a person who is skilled in the art.

9.4. Results

The analysis of the precedence was essential because it was used to support the opposition to the patent application filed in Brazil. Considering the Industrial Property Law no. 9 279/96, as well as the Normative Act 127/97[23], we can conclude that all the claims of the application PI9811045-4 do not involve any inventive step.

These conclusions were also noticed by "Doctors without Borders"-DWB, who were used to help Indian lawyers to adopt the same strategy in India[24].

The United States Patent and Trademark Office (USPTO) has granted four patents for Tenofovir drugs. The NGO Public Patent Foundation (PUBPAT) had requested the removal of these four patents in USA. After analysis, the patent examiner concluded to revoke all of them. However, after several months, the USPTO examiner changed his mind and he approved the four patents relating to Tenofovir.

The references listed in the opposition and in the technical report of the patent examiner show that bis (POC) PMPA was already known for the treatment of HIV/AIDS at the time of filing PI9811045-4 in Brazil. Thus, the

---

Emedastine Difumarate (Kanebo Ltd.), Quietapine Fumarate (ICI Americas Inc.), Ibutilide Fumarate (The Upjohn Company).

23 Normative Act no 127. Available at http://www.inpi.gov.br/legislacao/atos_normativos/ato_127_97.htm?tr2.

24 DWB supports the opposition for patent application by Gilead on tenofovir in India. Information available at: http://www.msf.be/fr/terrain/pays/asie/inde_news_13.shtml.

examiner noted that the conversion of bis (POC) PMPA in the Tenofovir disoproxil fumarate through the use of fumaric acid does not involve any inventive step, according to the references cited.

The INPI committee concluded that the application PI9811045-4 was in disagreement with the Articles 8, 10 (item VIII), 13, 24 and 25 of the LPI 9279/96. The technical report with the arguments was published on 8 April, 2008 in the RPI[25].

After the deadline, Gilead presented explanations to INPI, but the examiner did not change his mind and he decided to reject the patent application on 28 August, 2008, in Brazil.

This further underlines the significance of these results, thus reaffirming the necessity of establishing a network between the developing countries. The submition of a project about this context will enable us to provide a rationale base to adjust different countries' laws.

Studies, at a more theoretical stage, could provide a basis, visibility, and legitimacy to our scientific approach. These studies could explain, in particular, how the information contained in the databases could be used to create knowledge through a process widely described by Don Swanson in his book *Knowledge Discovery in Databases*. This process of opposition may be used in various processes of synthesis and formulation used in the development of molecules and drugs recommended in the AIDS treatment.

9.5. Conclusion

In view of the study that has been described in this chapter, we believe that a search for drug information (prior art, active ingredients, authorizations to commercialize a drug, state of international scientific and technical research, law) is a key necessity to active a measured cost reductions policy for the Brazilian state to maintain free and universal access to treatment for HIV. This research will be based on the knowledge and structures which have been implemented in Brazil already, especially in the context of the Department of STD, AIDS and Viral Hepatitis, and additional

---

25 RPI – *Journal of Industrial Property.*

skills to respond to the changing international context and, in particular, the implementation of TRIPS.

In fact, the legal framework about the introduction of compulsory licensing and parallel had imports appeared to be clear since 6 December, 2005 until the collapse of negotiations on the Doha Round on 24 July, 2006. So, it is therefore important that Brazil adopts all the elements necessary for it to be the engine for the developing countries as it has often been in the fight against AIDS and so as to take full advantage of this amendment if an agreement were to be signed between the WTO members.

The risks in the HIV treatment aren't only located on the medical plane. It is necessary to control the changing legal and regulatory environment, in terms of industrial protection and, in particular, it is vital to watch the regulatory environment that is not yet stabilized. The earliest possible information on these filed patents or drugs is a weapon that enables one to influence its environment.

So, it is necessary to implement a process of technological intelligence (patent monitoring) with focus on two issues:

– explain to the researchers and entrepreneurs the possibility of accessing a unique source of strategic information with data on research and production drugs, or drugs to be launched, in the treatment of AIDS and opportunistic diseases;

– explain to the government the importance to provide strategic information to the decision-maker.

This strategic information serves several actions, and provides the elements necessary to support arguments during the phases of negotiation with the patents assignee of ARVs and opportunistic diseases drugs; this information would help the policy-makers to identify opportunities for parallel importation of ARVs and drugs recommended for the treatment of opportunistic infections, and finally, demonstrate all the necessary arguments to formulate oppositions to the patent applications of ARVs at the INPI in Brazil, as per our recent study about Viread® drug where the result will be the possibility of producing the formulation and synthesis of ARVs on Brazil.

Other solutions that could reduce the costs with the treatment will be considered including the study of the strategy adopted by different patent applicants[26]. Finally, although the Brazilian capacity for producing ARVs seems significant, it doesn't concern the whole process of producing the drugs (which could be schematized into two main steps: synthesis and formulation). The Brazilian laboratories are "highly specialized in producing drug formulation" but have "a restricted capacity for producing active ingredients" in the composition of ARVs.

This chapter should complement the several studies already done about this subject in economics, being part of a North-South network, which is already established but is constantly developing to face the changing context on the Industrial Property Law (since the TRIPS amendment appears to be confirmed). Finally, it is part of the international cooperation with Brazil, where the concerned area has been accorded a priority by the Ministry of Foreign Affairs in France[27] and by the European Commission: Science and Information Technology and Communication.

This chapter should complement the studies that have been already carried out and this subject has been recognized as an economics subject, and is to be part of an already established North-South network, which is constantly developing to face a changing context in the Industrial Property Law at the international level (because the amendment of the TRIPS seems to be confirmed). Finally, it is part of an International Cooperation with Brazil, considered a priority by the Ministry of Foreign Affairs in France and by The European Commission: Science and Technologies of Information and Communication.

Lastly, considering that these research studies concern primarily the management of strategic information on drugs and their active ingredients, international information based on patent information, records of drugs in specialized agencies, scientific and technical publications, the obtained results will promote knowledge transfer in the South-South cooperation in the HIV treatment. This cooperation seems to be emerging with greater certainty as evidenced in:

---

26 INPI has launched a surveillance program of competitors from publicly available information on patents. Information available at http://www.fazenda.gov.br/resenhaeletronica/MostraMateria.asp?cod=268358.

27 Regional delegation for cooperation "France – Southern Cone", STIC Amsud, science and information technology and communications in South America. Available at http://www.france-conesud.cl/html/drc_STIC.htm.

– the creation of the Horizontal Technical Cooperation Group on HIV/AIDS [28];

– the establishment of technological cooperation network on HIV/AIDS among developing countries and emerging markets (South Africa, Brazil, China, India, Nigeria, Russia, Thailand, Ukraine)[29];

– the cooperation between Brazil and Latin America and the Caribbean (Bolivia, Colombia, El Salvador, Paraguay, Dominican Republic) and Portuguese-speaking countries (Cape Verde, Guinea-Bissau, Mozambique, Sao Tome e Principe and East Timor) under the International Cooperation Program with other developing countries[30], supported by UNAIDS and UNICEF.

9.6. Bibliography

[ABB 02] ABBOTT F.M., "The Doha declaration on the TRIPS agreement and public health: Lighting a dark corner at the WTO", *Journal of International Economic Law*, vol. 5, no. 2, p. 469-505, 2002.

[ANR 05] ANRS, Dossier de presse, December 1, 2005, available at: http://www. anrs.fr/index.php/filemanager/download/745/DP%201%20d%C3%A9c%2005% 20-%20 ANRS-RICAI.pdf.

[ANR 04] ANRS, Dossier de presse "Symposium Franco-Brésilien – Le nouveau droit de la propriété intellectuelle dans le domaine de la santé et du vivant", Brasília, 23-24 June 2004. Available at http://www.anrs.fr/index.php/ filemanager/download/457/Le%20nouveau%20droit%20de%20la%20propri%C 3%A9t%C3%A9%20intellectuelle%20dans%20le%20domaine%20de%20la%20 sant%C3%A9%20et%20du%20vivant.pdf.

[ANR 05] ANRS, Interview de Benjamin Coriat, December 2005. Available at http://www.cite-sciences.fr/francais/ala_cite/science_actualites/sitesactu/ magazine/article.php?id_mag=2&lang=fr&id_article=5383.

---

28 The Horizontal Technical Cooperation Group of Latin America and the Caribbean. Available at http://www.aids.gov.br/data/Pages/LUMIS8170CCBDITEMID2BD5044813 F94AB2812B7AC62B8584E1PTBRIE.htm.

29 A Experiência Inovadora da Rede de Cooperação Tecnológica em HIV entre 8 Países em Desenvolvimento. Available at http://www.aids.gov.br/data/Pages/LUMIS8170CCBDITEMID CEB3554AED13489289BBFE2ABD4B5065PTBRIE.htm.

30 Programa de Cooperação Internacional para Ações de Controle e Prevenção do HIV para Países em Desenvolvimento – PCI. Available at: http://www.aids.gov.br/data/Pages/ LUMIS8170CCBDITEMID9BAD2EFE0CB24C3B81BA870FAED5D2FFPTBRIE.htm.

[BAL 93] BALZARINI J., HOLY A., JINDRICH J., NAESENS L., SNOECK R., SCHOLS D., DE CLERCQ E., "Differential antiherpesvirus and antiretrovirus effects of the (S) and (R) enantiomers of acyclic nucleoside phosphonates: potent and selective in vitro and in vivo antiretrovirus activities of (R)-9-(2-phosphonomethoxypropyl)-2,6-diaminopurine", *Antimicrobial Agents and Chemotherapy*, no. 37, p. 332-338, February 1993.

[BEL 01] BELTRAME P.A., "Lei de Propriedade Industrial do Brasil em Discussão na OMC", *Revista da ABPI*, no. 55, 2001.

[BIS 96] BISCHOFBERGER N., CHE-CHUNG T., FOLLIS K.E., SABO A., GRANT R.F., BECK T.W., DAILEY P.J., BLACK R., "Antiviral efficacy of PMPA in macaques chronically infected with SIV", *Antiviral Research*, vol. 30, no. 1, p. 42-42(1), April 1996. Available at http://www.ingentaconnect.com/content/els/01663542/1996/00000030/00000001/art80307.

[BIS 97] BISCHOFBERGER N., NAESENS L., DE CLERCQ E., FRIDLAND A., SRINIVAS R.V., ROBBINS B.L., ARMILLI M., CUNDY K., CHOUNG K., LACY S., LEE W., JENGPYNG S., "Bis(POC)PMPA, an orally bioavailable prodrug of the antiretroviral agent PMPA", *4th Conference on Retroviruses & Opportunistic Infections*, Washington DC, January 1997, Abstract A463. Available at http://www.retroconference.org.

[BRA 96a] Brazilian Federal Law no. 9.279, of 14 May 1996. Available at http://www.inpi.gov.br/legislacao/leis/lei_9279_1996.htm.

[BRA 96b] Brazilian Federal Law no. 9.313, of 13 November 1996, called "Samey law", available at http://www.presidencia.gov.br/ccivil_03/LEIS/L9313.htm.

[CHE 95] CHE-CHUNG T., FOLLIS K.E., SABO A., BECK T.W., GRANT R.F., BISHOFBERGER N., BENVENISTE R.E., BLACK R., "Prevention of SIV infection in macaques by (R)-9-(2-phosphonylmethoxypropyl)adenine", *Science*, vol. 270, no. 5239, p. 1197-1199, 17 November 1995.

[CHE 05] CHEQUER P., "Access to treatment and prevention: Brazil and beyond", *3rd IAS Conference*, Rio de Janeiro, Brazil, 24-27 July 2005. Available at http://www.infectologia.org.br/anexos/IAS%202005_Plen%C3%A1ria%2025.07_Chequer_slides.pdf.

[COR 03] CORIAT B., ORSI F., "Brevets pharmaceutiques, génériques et santé publique. Le cas de l'accès aux traitements antirétroviraux", *La revue d'économie publique*, IDEP, Marseille, December 2003. Available at http://economiepublique.revues.org/sommaire65.html.

[EST 01] ESTRELA A., FLOH F., "A Lei de Patentes Brasileira e as Regras da Organização Mundial do Comércio", *Revista da ABPI*, no. 55, 2001.

[EUR 06] EUROPEAN COMMISSION, European Union seeks to strengthen partnership with Mercosur countries on information and communication technologies, Press release IP/06/91, of 30 January 2006. Available at http://europa.eu.int/rapid/pressReleasesAction.do?reference=IP/06/91.

[FRI 97] FRIDLAND A., ROBBINS B.L., SRINIVAS R.V., ARIMILI M., KIM C., BISCHOFBERGER N., "Antiretroviral activity and metabolism of bis(POC)PMPA, an oral bioavailable prodrug of PMPA", *Antiviral Research*, vol. 34, no. 2, p. 49-49(1), April 1997. Available at http://www.ingentaconnect.com/search/expand?pub=infobike://els/01663542/1997/00000034/00000002/art83167&unc.

[GRA 02] GRABOWSKI H., "Patents, innovation and access to new pharmaceuticals", *Journal of International Economic Law*, p. 849-860, Oxford University Press, Oxford, 2002.

[HAM 02] HAMMER P.F., "Differential pricing of essential AIDS drugs: Markets, politics and public health", *Journal of International Economic Law*, p. 883-912, Oxford University Press, Oxford, 2002.

[MFU 02] MFUKA C., "Accords ADPIC et brevets pharmaceutiques: le difficile accès des pays en développement aux médicaments antisida", *Revue d'Economie Industrielle*, no. 99, Les droits de la propriété intellectuelle: nouveaux domaines, nouveaux enjeux, numéro spécial, 2e trimestre 2002. Available at http://revel.unice.fr/reco/sommaire.html?id=34.

[MIN 05] MINISTÉRIO DA SAÚDE, A sustentabilidade do acesso universal a anti-retrovirais no Brasil, Brasília, 9 August 2005. Available at http://www.aids.gov.br/final/novidades/Cons_Nac_de%20Sa%FAde_pos_PN%20patentes.pdf.

[MOA 03] MOATTI J.P., CORIAT B., SOUTEYRAND Y., BARNETT T., DUMOULIN J., FLORI Y.A., *Economics of AIDS and Access to HIV/AIDS Care in Developing Countries, Issues and Challenges*, ANRS, Paris, June 2003. Available at http://www.anrs.fr/index. php/article/articleview/1114/1/317.

[NAE 97] NAESENS L., BISCHOFBERGER N., ARMILLI M., KIM C., DE CLERCQ E., "Anti-retrovirus activity and pharmacokinetics in mice of bis(POC)-PMPA, the bis(isopropy-loxycarbonyloxymethyl) oral prodrug of PMPA", *Antiviral Research*, vol. 34, no. 2, p. 50-50(1), April 1997.

[OMC 01a] OMC, "Déclaration sur l'accord sur les ADPIC et la santé publique", Conférence Ministérielle de l'OMC, Doha, Qatar, 2001. Available at http://www. wto.org/french/thewto_f/minist_f/min01_f/mindecl_trips_f.htm.

[OMC 01b] OMC, Les Membres examinent des projets de déclaration ministérielle, Conseil des ADPIC, 19 and 21 September 2001. Available at http://www.wto.org/french/news_f/news01_f/trips_drugs_010919_f.htm.

[OMC 03] OMC, Déclaration du président du conseil général de l'OMC of 30th August 2003. Available at http://www.wto.org/french/news_f/news03_f/trips_stat_28aug03 f.htm.

[OMC 05a] OMC, ADPIC. Available at http://www.wto.org/french/tratop_f/trips_f/trips_f.htm.

[OMC 05b] OMC, Approbation de l'amendement rendant permanente la flexibilité dans le domaine de la santé, Press release of 6 December 2005. Available at http://www.wto.org/french/news_f/pres05_f/pr426_f.htm.

[OMC 05c] OMC, Licences obligatoires pour les produits pharmaceutiques et Accord sur les ADPIC, October 2005. Available at http://www.wto.org/french/tratop_f/trips_f/public_health_faq_f.htm.

[OMC 05d] OMC, "Négociations, mise en œuvre et travaux du Conseil des ADPIC", *Conférence ministérielle de l'OMC*, Hong Kong, 2005. Available at http://www.wto.org/french/thewto_f/minist_f/min05_f/bricf_f/bricf06_f.htm.

[PIE 05] PIERRET J.D., DOLFI F., QUONIAM L., BOUTIN E., RICCIO E.L., "Découverte de connaissances dans les bases de données bibliographiques: modèles expérimentaux autour de la première hypothèse de Swanson", *Information Sciences for Decision Making*, no. 20, University of the South Toulon-Var, 2005. Available at http://isdm.univ-tln.fr/PDF/isdm20/pierret.pdf.

Chapter 10

Innovation, Serendipity 2.0, Filing Patents from Biomedical Literature Exploration

10.1. Introduction

Biomedical information professionals practice in an environment in which bibliographic databases are numerous and generally very well-structured, by using codes (molecules, genes, and proteins) and the thesaurus. The volume of information made available through these databases is considerably vast. The most popular of these databases is Medline, which is freely available through the PubMed Website[1]. Medline is produced by the National Library of Medicine under the Department of Health of the United States. It has currently 17.2 million references among which more than 528,000 were added in 2008. Embase[2], produced by Elsevier, offers access to over 12 million references.

The volume and rate of growth of information available can be analyzed from two perspectives. On the one hand, we can consider that information is better disseminated, so that these databases have a coverage which tends toward some exhaustiveness, especially if they are combined in the same bibliographic search. But on the other hand, the information sought may be

Chapter written by Jean-Dominique Pierret and Fabrizio Dolfi.
1 http://www.pubmed.gov [page consulted on March 17, 2009].
2 http://www.elsevier.com [page consulted on February 4, 2009].

difficult to obtain, either because of the size of such databases (needle in a haystack [GRI 02]), or because of knowledge fragmentation.

The latter concept requires some explanation. At present, our knowledge increases constantly and scientific literature increases accordingly. In order to process and upgrade the growing volume, the scientific community seems to organize itself by sharing the work, therefore identifying more precise problems by creating increasingly sophisticated disciplines and specialties [SWA 93]. Thus, the volume of information on these new specialties is nearly constant and manageable by specialists.

Gradually, as the literature grows, the number of specialties increases, leaving to each of them an approximately stable number of publications. With the ratio between the number of researchers and literature remaining constant, this allows everyone to follow – or leaves everyone with the illusion of following – the literature in his/her field. The consequence of this increase in specialities is the fragmentation of knowledge.

For Morin, "the principle of separability has emerged in scientific domain through specialization, then it deteriorated into hyper-specialization and compartmentalization" [LEM 99]. Therefore, with the dividing of scientific production, the logical relationships between different specialties tend to be overlooked, ignored, or hidden.

However, two specialties can be complementary in some respect and together, can be carriers of a new knowledge, and even if scientists were organized to follow up with what is produced in their fields of expertise – becoming highly restricted – it is now impossible for one man to grasp all the knowledge in a field such as molecular biology or quantum physics. The remoteness of disciplines which can be complementary in some respects and the increasing volume of available information help increase the number of connections that are carriers of the new knowledge.

The concept of "knowledge discovery in bibliographic databases" (that will be henceforth referred to in this chapter as LBD – abbreviation for Literature Based Discovery) is to reveal the unsuspected connections. It contains a paradox inherent in the information source used and breaks from traditional methodologies of information retrieval.

Information available in bibliographic databases is dated information, validated by a long process that makes it less innovative. How can we, with this restriction, bring out innovative solutions through this information?

The solution is probably to be sought in deployed methodologies which are different from the classical querying of databases using Boolean logic. When searching for information in a database in a conventional manner, the user sends a query to the database and only gets back information that is already published.

The query result is a set of known information that does not possess any newness. The use of Boolean operators does not enable us to move toward updating new knowledge.

Thus, the use of "AND" highlights elements common to two themes (fish oil and Raynaud's disease) or – at best – lack of a common element, which in the first work of Swanson reveals that his hypothesis is not supported by any direct link and may perhaps be the source of a new discovery.

In LBD literature, it is not the presence of a response to a query that is significant, but rather the lack of response. This is therefore a different paradigm, as compared with the information retrieval.

Knowledge fragmentation and increase in the volume of information are responsible for hiding these relevant relationships. The work of Don Swanson, the foundation of our LBD method, focuses on the identification of such relationships through the exploitation of bibliographic databases. The term used by Swanson was undiscovered public knowledge [SWA 86].

As scientific production increases at the same time as scientists specialize in highly complex areas, there would be hidden connections. Swanson's method adapts well to this explosion of information and it takes full advantage of it.

This may seem paradoxical. On the one hand, the vast majority of users of information sources have to face the task of finding relevant information which is difficult due to the increased volume of information, while on the other hand, the method of Swanson takes advantage of this wealth and suggests to those who exploit it, new opportunities for discovery.

10.2. The work of Don Swanson

10.2.1. *Fish oil and Raynaud's disease*

Don R. Swanson, a physicist by training, has demonstrated a great interest for biomedical information throughout his career. He is now professor Emeritus at the University of Chicago. In the early 1980s, Don Swanson noticed an article on the Eskimo diet. Consumption of fish and marine mammals (which are rich in long polyunsaturated fatty acids) reduces the risk factor for cardiovascular disease, and hence their lower incidence in Eskimos [DYE 82]. Swanson then performed a series of literature searches in this direction and he found that:

– fish oil, composed largely of such fatty acids, was known to reduce blood viscosity and platelet aggregation (helps prevent thrombosis and atherosclerosis) and to act on vascular reactivity;

– in Raynaud's disease, blood viscosity and platelet aggregation increase and an exaggerated vasoconstriction occurs.

The link is obvious and Swanson is the first to formulate the hypothesis that fish oil is a potential treatment for Raynaud's disease. In fact, before 1986, there was no document that linked fish oil and Raynaud's disease. One of his publications detailed his hypothesis from a physiological point of view [SWA 86] and another outlined the method used [SWA 87]. In 1989, a team of clinicians from Albany Medical College in New York demonstrated that even if fish oil cannot cure Raynaud's disease, it helps to improve the condition of patients [DIG 89]. Swanson focused on fish oil as he was interested in the role of diet on health: the impact of dietary factors on the decline of the disease.

Swanson summarized the background of his discovery: "In 1985, I was struck by lightning and have never recovered" [SWA 01]. He realized that two items of information from different medical articles suggest, when they are juxtaposed, a hypothesis that nobody knew then. Connecting two disjoint pieces of information can create new knowledge. His approach was more intuitive than formal. At the same time, he regretted that he was not able to describe the systematic process of finding hidden connections [SWA 87]. However, he quickly formulated a strategy based on the use of Medline, Embase and SciSearch bibliographic databases, dubbed explore/exclude or trial-and-error. This strategy would allow us to search for connections between two non-interactive and non-complementary articles, (not citing each other) in order to generate information missing from the two articles, if considered separately [SWA 89]. His work focuses mainly on improving his methodology and the discovery of new hypotheses.

10.2.2. *Other assumptions by Swanson*

Since 1986, Swanson has published over 25 articles on the subject and offered no fewer than 8 other assumptions including the identification of potential links between:

– migraine and magnesium [SWA 88];

– phospholipase A2 and schizophrenia [SMA 98a];

– estrogen and Alzheimer's disease [SMA 96a];

– indomethacin and Alzheimer's disease [SMA 96b].

10.2.3. *Modeling Swanson's methodology*

We will use the term "ABC model" for convenience. Swanson drafted the following thought process:

– A, fish oil;

– B, platelet aggregation and blood viscosity;

– C, Raynaud's disease.

A improves C by acting on B. This is the classical pattern of drug action. Disease C is characterized by a number of physiological disorders B (pathophysiology), Drug A acts favorably on physiological disorders. The revelation of hidden associations may follow an open or close process.

Hypothesis generation may follow an open process. The starting point is the known literature A, the aim being to identify B and C, where B and C are unknown, *a priori*. Other variations are also possible e.g. C → B → A or A ← B → C. The exploratory phase of this strategy is an open process: starting from Raynaud's disease, Swanson looks for physiological disorders involved in the disease, and then identifies a potential treatment for an unknown *a priori* [SWA 89].

A closed process tests a hypothesis, identifies the relationship between A and C: what are the links between migraine and magnesium [SWA 88]? From a general point of view, testing a hypothesis entails working on a more limited amount of information than in generating a hypothesis.

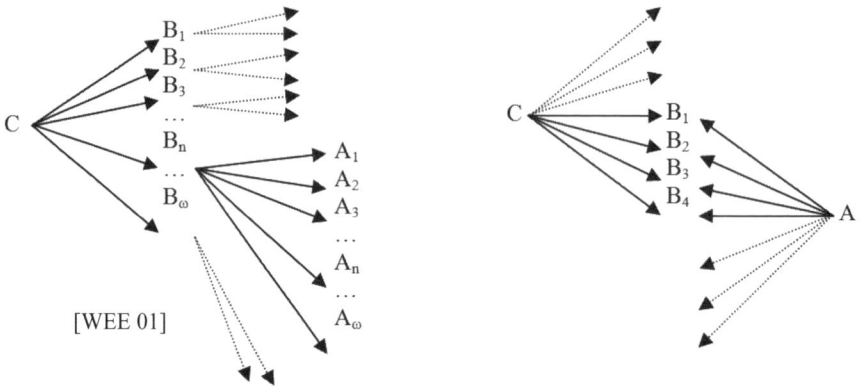

Figure 10.1. *Open and closed process*

10.2.4. *Tools from the Swanson methodology*

Swanson's model was used to support the creation of several LBD derivatives tools, alternately combining the expertise and automatic processing of large volumes of information. Two examples of this are Arrowsmith [SMA 98b] and DAD [WEE 00, WEE 01, WEE 03]. Other teams have developed their own systems [BLA 02, DEM 03, GOR 96, HRI 01, PER 04, SRI 04, STE 03, WRE 04, YET 06].

In the 1990s, with the help of Neil Smalheiser, Don Swanson developed Arrowsmith, a computerized system which was freely accessible through the Internet to explore possible links between two citations files from Medline (validation of hypothesis). Arrowsmith is designed to guide the user in the discovery of implicit relationships by helping to select relevant bibliographic references to establish relationships AB and BC. Swanson and Smalheiser used Arrowsmith to replicate many of the works of Swanson [SMA 96b, SMA 98a, SMA 98b, SWA 99].

The method that we have developed is based on Swanson's transitive reasoning. It combines treatment of bibliographic information, analysis and serendipity. Before describing it, it is important to note that the unexpected forms the basis of the obtained results. LBD should allow us, after reflection, to generate new hypotheses. It is a creative process. One should therefore be open and agree to be surprised by the unexpected, have the ability to think otherwise, apart from the usual frameworks [SWI 05]. At the operating level, the method we had deployed combined bibliometric data treatment achieved in a semi-automatic way with the validation of data by the expert whose role is decisive. Cognitive biases are at work to avoid assumptions not confirmatory of previous ideas. The expert must instead be open to all possibilities.

10.3. Diseases-Physiopathology-Molecules (DPM)

10.3.1. *A new mode of transition*

Exploring the literature in search of signs indicating the pathophysiology[3] of a disease is, in the context of biomedical research, usually an unnecessary step. Indeed, it is often a team work where expertise on the disease is available. We therefore propose to solicit directly the experts and ask them to define which physiological disturbances are involved in the disease of interest.

Thus, replicating the approach of Swanson, the first question we will ask is: what are the physiological disturbances brought into play in Raynaud's disease? Considering the state of knowledge in November 1985, when Swanson wrote his first paper, it is obvious to the specialist that Raynaud's disease involves:

– high blood viscosity;

– an increase in platelet aggregation;

– vasoconstriction of vessels supplying the fingers;

– greater rigidity of erythrocytes, red blood cells.

Searching for the molecules that can act on each of these phenomena, independently of each other, can lead to finding new ways to treat Raynaud's disease. Thus, by describing the pathology in pathophysiologic phenomena and only being interested in these phenomena and not the disease, the problem is moved to a larger field of course, and can also offer an original approach in terms of treatment.

---

3 Physiology is the science which studies the function of living things or their organs, physical or chemical factors and processes involved. Pathophysiology concerns the modifications of major physiological functions during disease.

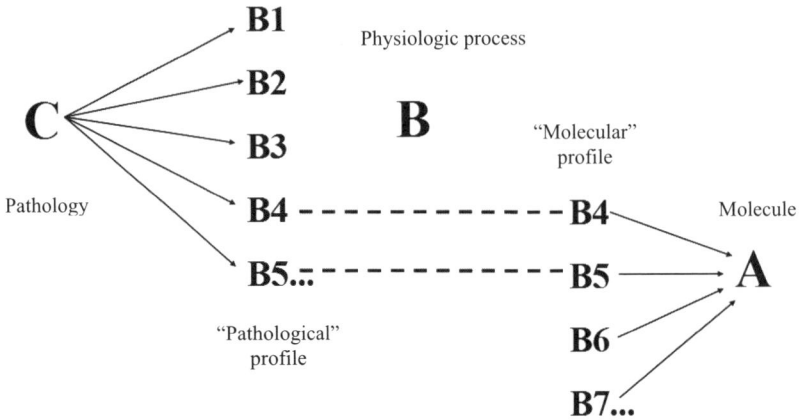

Figure 10.2. *ABC model, adapted to DPM approach*

In this particular example, irrespective of the fact that one or more of these four aspects may be considered irrelevant today, Swanson selected them, based on the state of knowledge in 1985. For the remainder of this chapter, we will keep to the conditions close to those of Swanson, taking into account the knowledge at the time, which might be called the truth of the time.

The diagram in Figure 10.2 can be modified and extended to various kinds of links between a disease and a molecule, as long as they are logical. For example, instead of observing the physiology, it is quite possible to look for proteins whose disturbed regulation may characterize certain diseases on which molecules can act. The search for co-occurring or adjacent diseases in a given disease is also possible. This process can also be used to identify not only a molecule, but also a protein or gene. It can also be used to re-address a molecule: by describing all the physiological targets of this molecule, we will be able to ascertain the conditions that are shared by one or more of these targets. Everything is possible provided that the logic of reasoning by transitivity is complied with i.e. that the links make sense.

DPM is based on two pillars: one technical, the realization of non-Boolean searches to extract common concepts, the other pillar being human. The latter is critical. The data analysis must be done in a spirit of openness, accepting to be surprised. Serendipity is required; it entails observing and capturing the surprising facts. Curiosity is not a bad thing. This mindset must be based on solid and extensive knowledge of the biomedical field (general knowledge) and the ability to accept surprises. The interpretation of these observations follows a mode of reasoning by abduction; that is to say, explaining the data observed in a broader framework of knowledge. The data is generated to produce the unexpected, to discover and understand. "In observation sciences, chance only favors the prepared mind" said Pasteur. Ideally, teams that employ such methods must have the following skills:

– expertise in information science, with extensive knowledge of sources and associated tools;

– expertise in the biomedical field;

– the culture of listening, sharing information and the ability to go beyond the usual reasoning frameworks.

10.3.2. *Non-Boolean treatment of information*

We therefore propose to take the example of the first discovery of Swanson to illustrate the functioning of the DPM. The bibliographic corpus processed is prior to November 1985: the date Swanson wrote his article [SWA 86].

The bibliographic references from Medline are indexed by using the MeSH (Medical Subject Headings): a medical thesaurus with its 2009 edition including 25,186 descriptors. MeSH is available online[4] and our work was based on this version. One of the features of MeSH is to designate major and secondary descriptors for a given reference. Major descriptors highlight key themes of a reference. In addition, each descriptor can be weighted by one or more qualifiers (there are 83), which will clarify the context in which the keyword is used. For example:

– aspirin/Therapeutic usage means that the reference addresses the use of aspirin for treating a disease;

– pancreas/physiology, the reference addresses the physiology of the pancreas.

---

4 http://www.nlm.nih.gov/mesh/meshhome.html, page consulted on 6 February 2009.

Combining qualifiers and major or non-major descriptors, it is possible to precisely query Medline. The answer to the question "what are the physiological disturbances brought into play in Raynaud's disease?" can be translated into MeSH descriptors in different ways. To illustrate our purpose, we selected the following:

– Erythrocyte Deformability [Majr] Limits: Enter Date to 1985/11: 135 references;

– Blood Viscosity [Majr] Limits: Enter Date to 1985/11: 1876 references;

– Muscle, Smooth, Vascular [Majr] or Vasoconstriction [Majr] Limits: Enter Date to 1985/11: 3864 references;

– Platelet Aggregation [Majr] Limits: Enter Date to 1985/11 to: 4567 references.

[Majr] means that the descriptors sought are major. These queries were executed on February 6, 2009 by using PubMed. They allow us to obtain four literatures. The issue here is no more that of Raynaud's disease, but that of four distinct physiological phenomena.

For each application, the references were downloaded to a hard disk in Medline format, which allows visualizing the descriptors used for each reference. Each of the four files is then reformatted to keep the descriptors only. All other bibliographic fields, the label of the descriptors field, where "*" sign implies that a descriptor is major, as well as qualifiers were eliminated. This work can be done by using word processing software furnished with macro commands, or using Basic or Perl routines. We got four lists of descriptors, one for each literature.

We also downloaded MeSH from the NLM site, in the form of MeSH Tree, which only contains the descriptors and their codes in a single file. The editing of this file allows selecting descriptors for Dietary Factors which Swanson had studied. Electrolytes, elements, isotopes, dietary carbohydrates and lipids, which form a list of 516 descriptors, have been retained.

The operation that followed was to extract dietary factor descriptors common to the four literatures. Again, the Basic or Perl routines can be created. We can also take advantage of relational database management software by loading the four literatures and dietary factors on five tables, and by cascading two queries: one extracts the common descriptors of the four literatures and calculates their frequencies, with the second retaining only the dietary factors from the first literature. The result is presented in tabular form and is handled with a spreadsheet.

The diagram in Figure 10.3 illustrates the collection and processing of the four literatures. Other ways of treating the subject could be considered: the definition of descriptors to search Medline; the use or non-use of major descriptors, major qualifiers or MeSH hierarchy for querying; the choice of dietary factors, the querying of other bibliographic fields (title, abstract or other codes), the extraction of descriptors of each literature (major or all, with or without particular qualifiers). It is an iterative process of trial and error, which must lead to obtaining some relatively short lists (some tens of words) with minimum noise. The noise is estimated by attempting to analyze a small number of concepts from the list: if they do not lead to any track, do not stimulate any reflection, even indirect, we believe that state implies the presence of noise – or that we are not able to understand the nature of links that support these terms. In such a case, the test proves to be an impasse and, most of the time, terms used to query Medline should be modified.

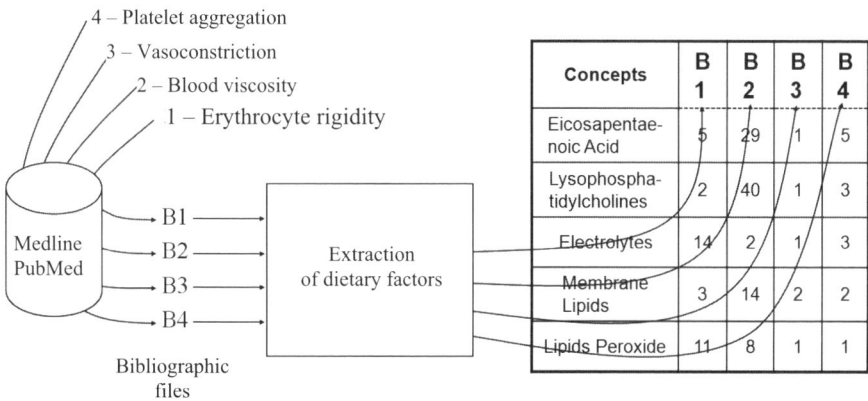

Concepts	B1	B2	B3	B4
Eicosapentae-noic Acid	5	29	1	5
Lysophospha-tidylcholines	2	40	1	3
Electrolytes	14	2	1	3
Membrane Lipids	3	14	2	2
Lipids Peroxide	11	8	1	1

4 – Platelet aggregation
3 – Vasoconstriction
2 – Blood viscosity
1 – Erythrocyte rigidity

Medline PubMed
B1
B2
B3
B4
Bibliographic files

Extraction of dietary factors

Figure 10.3. *DPM schema replicating the first experiment of Swanson*

As summarized in the diagram in Figure 10.3, the four literatures are processed in parallel, either for collection or extraction of concepts of interest (here, dietary factors). They were not combined until the final stage, and the result of selection of common concepts is summarized in the table.

10.3.3. *Analysis of results*

The final result is a table (Table 10.1) containing the list of 21 dietary factors common to the four literatures. The terms marked with an asterisk * are those that have caught our attention. EPA (eicosapentaenoic acid) is a component of fish oil.

From this table, the expert in the biomedical field will be able to test each potential hypothesis. Taking the example of EPA, the first step is to consult the literature and see the impact that EPA has on the deformability of erythrocytes, platelet aggregation, vasoconstriction, and blood viscosity.

MeSH dietary factors	Erythrocyte deformability	Platelet aggregation	Vasoconstriction	Blood viscosity
Calcium	5	189	648	9
Cholesterol	3	98	60	52
Oxygen	5	16	115	47
Epoprosteno	1	410	155	2
Potassium	4	8	371	7
Lipids	1	68	13	42
Sodium	3	9	139	14
Phospholipids	4	65	16	8
Triglycerides	1	63	13	26
Unsaturated fatty acids	1	62	8	4
Fatty acids	1	46	8	4
Eicosapentaenoic acid*	1	29	5	5
Lysophosphatidylcholines	1	40	2	3
Membrane Lipids	2	14	3	2
Electrolytes	1	2	14	4
Radioisotopes	1	18	7	1
Fish Oil*	1	12	2	2
Lipid Peroxides	1	8	11	1
Lithium	1	8	6	1
Phosphates	1	3	6	1
Phosphorus	1	3	1	1

Table 10.1. *Results of DPM on Raynaud's disease*

If the EPA action on each of these four physiological phenomena is consistent with an improvement of Raynaud's disease, the EPA hypothesis should then be further studied by further reviewing the literature (journals and reference works). If despite such a study, the proposed hypothesis still holds, then the final step before accepting or rejecting it will be the carrying out of experimentation.

Due to the systematic return to literature offered by the association of words in rows and columns, there are various ways to exploit such a table. Thus, a row can carry a hypothesis or the presence of converging elements in several rows, identified through the sagacity of the expert, can also lead to a hypothesis. Finally, some rows or even the majority will be no source of inspiration. We thus summarize the DPM cycle as containing four parts:

– choice of pathology for analysis, definition of pathophysiological phenomena to take into account;

– translation of the pathophysiology into Medline queries and querying of Medline;

– concept extraction and presentation in tabular form;

– analysis by experts and return to literature.

Roughly, the time needed to complete the first three steps is one day, provided that one possesses a clear representation of pathophysiological mechanisms involved in the disease studied. The fourth step is the longest, since it alternates analysis of results by the expert, information retrieval and generation of new DPM tables following the amendment of certain parameters. Progress is made through successive steps and by trial and error. Several days are required to propose a supported hypothesis, which can then be submitted to experimental verdict.

10.3.4. *DPM bias*

Like any other method, the induced biases in DPM should be known in order to better assess the value of results in the tables. Four of them are presented below:

– the nature of the relationship between two concepts: a link is supposed to exist between two concepts if they are collocates. However, both concepts may well be in the same article without being connected to each other in a logical manner. Analysis of the concepts in the tables should take this aspect into consideration as this is certainly the main source of false positives;

– the use of MeSH can induce three kinds of bias. First, the MeSH has certainly more than 25,000 words, but does not describe all the concepts involved in the biomedical domain: a huge domain where molecular, cellular, organs and organizms levels are closely dependent, involving relationships and interactions of a great complexity. Many aspects of biology are not precisely described by the MeSH concept and there is also the issue of novelty. All controlled vocabulary or thesaurus integrates well established concepts in the concerned discipline, leaving innovation and recent developments aside until they are stabilized, confirmed and incorporated. A gap results between new concepts and those of a thesaurus. Finally, the creation of dictionaries from the MeSH itself may introduce a bias: DPM will only retrieve concepts included in dictionaries used for a given analysis;

– the choice of physiological phenomena: the contents of a DPM table are entirely dependent on the choice of physiological phenomena to be treated, in quantity and quality. The more a DPM analyzes physiological phenomena simultaneously, the less it will highlight common elements. The quality of a DPM table also depends on the choice of descriptors: too general, the produced table will be long with a lot of noise; too restrictive, the table may be empty. Conducting various iterations around a particular topic, by changing the number of physiological phenomena studied, their combination and their formulations into concepts are necessary to produce a DPM analysis;

– thesaurus and negative results: MeSH terms describe the concepts discussed in an article, but they do not enable describing the nature of the results. Both proposals "A cure C" and "A do not treat C" will be described in the same way, with the same descriptors.

10.4. Conclusion: the place of LBD today

10.4.1. *Development of drugs and LBD*

The marketing of a new drug is a long, risky and expensive process, and takes place in a highly competitive environment: from the discovery of chemical components to the production of a drug takes an average of 14 years, compounds having a 0.0001 probability of being sold with development costs ranging between $500 million to $2 billion [ADA 09]. The pharmaceutical industry regularly goes through phases of restructuring, with mergers between laboratories in search of synergy: portfolios' complementarities of products in R&D, synergies in terms of marketed products or geographic coverage. Thus, since the beginning of the year, three merger operations have taken place: Pfizer (USA) acquired Wyeth (U.S.) for $68 billion thereby emerging as the world leader, Merck & Co (USA) Schering-Plough acquired (United States) for $41.1 billion, leading to the formation of the second largest pharmaceutical company, and finally, Roche (Switzerland) acquired Genentech (U.S.) for $46.8 billion.

The creation of new drugs is the result of a long process, punctuated by multiple steps including biology, chemistry, pharmaceutics and clinical experimentation. Very large amounts of data are generated and data-mining techniques are often used, particularly in the early stages of research [ADA 09]. It is in this context that LBD methods were evoked. However, feedbacks are very rare, if not nonexistent. We had the opportunity of discussing on two occasions with representatives from two of the top ten laboratories worldwide. These people work in bioinformatics on the one hand and the other hand in information service. Without going into details, we will highlight the two different testimonies we collected:

– the department of bioinformatics created a LBD system that allowed researchers to propose testable hypotheses from a clinical standpoint i.e. in humans. Biologists have not necessarily welcomed this tool with great enthusiasm; however those involved in clinical research are interested in it and made use of it. We were told that studies have been conducted and some patents have been filed;

– the LBD system was presented as part of the range of tools available to users. Information obtained from our source reveals that this is an abandoned project – or almost – since the key players were on both sides of the Atlantic.

LBD can be a useful tool for the development of drugs. Simply, two main types of application can be considered:

– identifying new therapeutic targets and new molecules;

– repositioning of molecules or known drugs for new therapeutic indication.

Repositioning strategy is being used by the pharmaceutical industry. This provides a new life, by opening new markets for a product already known, so that the risk of failures related to the toxicity of the product is limited [CHO 07]. Sildenafil (Viagra®) was developed for angina pectoris and then repositioned for erectile dysfunction or Botox, which was used to treat spasm associated with certain dystonias repositioned in the wrinkles. These are two well known examples. It must be also noted that serendipity is often involved in drug repositioning [SWI 05].

In this framework we applied DPM; our work has led to the filing of a series of patents [DOL 05, PIE 07a, PIE 07b, PIE 07c, TRE 07].

10.4.2. *The fate of the work of Swanson*

Swanson's work was welcomed in different ways, depending on whether one takes the side of information science or of natural science. Spasser published a citation analysis of Swanson articles to monitor how his ideas were perceived by the scientific community [SPA 97]. He thus identified 21 articles in information science and 12 in natural sciences which have cited at least one paper by Don Swanson on the LBD. In information science, the authors consider the method and techniques used; the generated medical hypotheses were taken as illustrations by them. For them, the existence of hidden public knowledge is a fact, like the ability to generate new knowledge through the clever use of information systems.

In natural science, Spasser shows that, expectedly, the authors were interested in the formulated hypothesis rather than the method. Their speeches on the hypothesis were rather negative and condescending, although they used it to buttress their arguments. In this case, Swanson's hypotheses are awaiting actual scientific validation. According to Spasser, the standards of the traditional biomedical research are at odds with the LBD. The purpose of scientific research is to predict and control, and not to understand the result of serendipity; it is oriented toward verification rather than discovery; his approach is focused on the specifics instead of being open to an apprehension of global problems or questions. He sees nothing surprising, in that researchers in natural sciences are reluctant to consider an exploratory method, which cannot be tested empirically or quantitative. This approach, often negative in the field of natural science, is a real cultural change.

The application of Swanson's method revolutionizes knowledge discovery methods. For example, it can greatly accelerate the stages of development of a new drug by being freed from lengthy and expensive phases. It is not surprising therefore that the people in charge of this process are quite critical of this method. In some cases, the works of Swanson are not an object of negative rhetoric: the authors use Swanson's articles to mention a relationship or a fact established empirically. Concluding his paper, Spasser stresses that while it is useful in biomedical research and particularly interesting, the concept of hidden public knowledge as well as the method of highlighting them remain largely unknown to researchers. Accordingly, this concept is itself a hidden public knowledge.

However, in the small world of research on LBD, the works of Don Swanson, particularly his hypotheses on Raynaud's disease, are considered as standards.

The penetration of LBD among its potential users seems non-existent today. This can be ascribed to several reasons:

– tools are – mostly – too heavy to implement (DPM operates from a standard office equipment);

– the thinking mode of researchers in biology does not accommodate the one proposed by the LBD as it has been suggested by Spasser;

– those who have implemented LBD use it successfully and, why not, want to shut this practice up to maintain a competitive advantage;

– there is no standard that allows evaluating the hypotheses generated by LBD.

On this last point, Kostoff said that in the LBD domain applied to natural sciences, the starting point is Don Swanson [KOS 08]. Virtually all works carried out have replicated the generation of the hypothesis on Raynaud's disease. However, according to Kostoff, it is not clear that this assumption is really original, since, he argues that the person skilled in the art could implicitly have knowledge of the usefulness of fish oil to relieve patients with Raynaud's disease, long before Don Swanson published his first article on the subject. To support his reasoning, he explained that in the 1970s, some articles dealt with the use of fish oil for vascular disease, and more specifically focused on peripheral vascular disease in the early 1980s, including Raynaud's disease. Kostoff emphasizes the robustness of control standard of hypotheses in order to ensure their originality, lack of prior art, and of course their validity.

The fact remains that the evaluation of LBD systems by testing hypotheses generated by them, is non-existent today. To our knowledge, except for the first hypothesis of Don Swanson, the fish oil and Raynaud's disease, no other hypothesis generated by LBD has undergone an experimental test. Under such conditions, how can we evaluate the results? How do we know whether they will stand experimentation? Indeed, by definition, these systems help to formulate new hypotheses that must then be proved. However, to overcome this difficulty, many LBD researchers have replicated the works of Don Swanson. They then considered that their systems function, when they produced the same results as those obtained by Swanson. From there, they sometimes offered few original hypotheses. Don Swanson did not limit his research to Raynaud's disease alone, but has also proposed 8 other hypotheses. However, for over 20 years, no major clinical trial has tested hypotheses generated by LBD. Rigorous evaluation of the LBD may be a barrier to its diffusion [YET 08]. The evaluation should not only lead to improving the effectiveness of LBD tools, but should also help in better integrating them into the daily practice of researchers – by calling the information specialist into the saddle in some cases – so that they can be useful.

Don Swanson, aware of this critical point, suggested that such results can – at least partly – be recognized as successful, if they are published in refereed journals [SWA 93]. In the same vein, we believe that if a hypothesis generated by a system of LBD can lead to a commercially exploitable patent, this hypothesis can then be said to be valid.

The main difficulty encountered by researchers who were interested in LBD is that their ambition should go beyond the extraction of explicit relationships within a document (as does the text-mining). The ambition of LBD is to identify implicit relationships within a corpus of documents which will have the following qualities: be real, original, new and non-trivial [SMA 08]. However, researchers who develop these methods do not have access to tools to test their hypotheses, as in natural science fields: vitro pharmacology tools or clinical studies in humans. Besides the "interdisciplinary distance" between natural science laboratories and information science laboratories[5], getting necessary funds to verify the

---

5 The case of the Psychiatric Institute of the University of Illinois at Chicago remains an isolated example of link between the two disciplines (neuroscience and medical informatics in this case). This is the institute to which N. Smalheiser and his team belong: http://arrowsmith. psych.uic.edu, accessed March 16, 2009.

results obtained by a method of knowledge discovery is a challenge. We must accept the reality that it is now virtually impossible to know whether a hypothesis is experimentally valid.

The future of LBD can be imagined as one of the possible multiple options between two ways. On the one hand, the further development of existing tools, with emphasis on results' validation, for example, by managing such projects within multidisciplinary teams (information science and natural science) which have the means to test hypotheses. And, secondly, the development of information retrieval tools based on LBD, for large number of users which could be integrated into general or specialized search engines, and also particularly oriented toward the end user; it would be new arrangements for access to information, which could form part of the information literacy programs advocated by some national and international organizations. Continuing in this way, Smalheiser proposes to assess the success of LBD by observing in the future its daily use by researchers – those at the bench top [SMA 08]. Will LBD be exploited on a large scale in the future without the need to understand its operation in order to benefit from it? This is a bit like Google which is widely used to explore the Internet but only few people know how it works. Google nevertheless remains a useful tool for researchers.

Swanson always considered that the purpose of a LBD system is to stimulate human creativity to produce plausible and verifiable hypotheses made in such a way that they can be published, criticized and in turn used as stimuli for further work [SWA 08].

In memory of Ugo

10.5. Acknowledgments

We thank Luc Quoniam for giving us the opportunity to contribute to this work and for the discussions we had together on this topic. We also thank Eric Boutin for his support and encouragement during the long period of development of the DPM.

10.6. Bibliography

[ADA 09] ADAMS C.P., BRANTNER V.V., "Spending on new drug development", *Health Economics*, 2009.

[BLA 02] BLAKE C., PRATT W., "Intervening in the life cycles of scientific knowledge", in *AAAI Spring Symposium on Mining Answers from Texts and Knowledge Bases*, Palo Alto, California, 2002.

[CHO 07] CHONG C.R., SULLIVAN D.J. JR., "New uses for old drugs", *Nature*, vol. 448, no. 7154, p. 645-646, August 2007.

[DEM 03] DEMAINE J., MARTIN J., DE BRUIIJN B., "Haystacks and hypotheses", in *ASIST 2003 Annual Meeting – Humanizing Information Technology: From Ideas to Bits and Back*, Westin Long Beach, California, 2003.

[DIG 89] DIGIACOMO R.A., KREMER J.M., SHAH D.M., "Fish-oil dietary supplementation in patients with Raynaud's phenomenon: A double-blind, controlled, prospective study", *American Journal of Medicine*, vol. 86, no. 2, p. 158-164, February 1989.

[DOL 05] DOLFI F., TREMEL N., Use of ondansetron for the treatment of inflammation, and pharmaceutical compositions thereof, European patent no. EP1600158, 2005.

[DYE 82] DYERBERG J., JORGENSEN K.A., "Marine oils and thrombogenesis", *Progress in Lipid Research*, vol. 21, no. 4, p. 255-269, 1982.

[GOR 96] GORDON M.D., LINDSAY R.K. "Toward discovery support systems: A replication, re-examination, and extension of Swanson's work on literature-based discovery of a connection between Raynaud's and fish oil", *Journal of the American Society for Information Science*, vol. 47, no. 2, p. 116-128, 1996.

[GRI 02] GRIVELL L., "Mining the bibliome: Searching for a needle in a haystack?", *EMBO Reports*, vol. 3, no. 3, p. 200-203, April 2002.

[HRI 01] HRISTOVSKI D., STARE J., PETERLIN B., DZEROSKI S., "Supporting discovery in medicine by association rule mining in medline and UMLS", *Studies in Health Technology and Informatics*, vol. 84 (Pt2), p. 1344-1348, 2001.

[KOS 08] KOSTOFF R.N., "Where is the discovery in literature-based discovery?", *Literature-based Discovery*, Springer-Verlag, Berlin, Heidelberg, 2008.

[LEM 99] LE MOIGNE J.L., *L'intelligence de la complexité*, L'Harmattan, Paris, 1999.

[PER 04] PERSIDIS A., DEFTEREOS S., PERSIDIS A. "Systems literature analysis", *Pharmacogenomics*, vol. 5, no. 7, p. 943-947, October 2004.

[PIE 07a] PIERRET J.D., DOLFI F., LOESCHE C., TREMEL N., Use of azasetron for the treatment of rosacea, and pharmaceutical compositions, international patent no. WO2007138234, 2007.

[PIE 07b] PIERRET J.D., DOLFI F., LOESCHE C., TREMEL N., Use of zatosetron for the treatment of rosacea, and pharmaceutical compositions, international patent no. WO2007138233, 2007.

[PIE 07c] PIERRET J.D., DOLFI F., LOESCHE C., TREMEL N., Use of granisetron for the treatment of sub-types of rosacea, and pharmaceutical compositions, international patent no. WO2007138232, 2007.

[SMA 96a] SMALHEISER N.R., SWANSON D.R., "Indomethacin and Alzheimer's disease", *Neurology*, vol. 46, no. 2, p. 583, February 1996.

[SMA 96b] SMALHEISER N.R., SWANSON D.R., "Linking estrogen to Alzheimer's disease: An informatics approach", *Neurology*, vol. 47, no. 3, p. 809-810, September 1996.

[SMA 98a] SMALHEISER N.R., SWANSON D.R., "Calcium-independent phospholipase A2 and schizophrenia", *Archives of General Psychiatry*, vol. 55, no. 8, p. 752-753, August 1998.

[SMA 98b] SMALHEISER N.R., SWANSON D.R., "Using ARROWSMITH: A computer-assisted approach to formulating and assessing scientific hypotheses", *Computer Methods and Programs in Biomedicine*, vol. 57, no. 3, p. 149-153, November 1998.

[SMA 08] SMALHEISER N.R., TORVIK V.I., "The place of literature-based discovery in contemporary scientific practice", *Literature-based Discovery*, Springer-Verlag, Berlin, Heidelberg, 2008.

[SPA 97] SPASSER M.A., "The enacted fate of undiscovery public knowledge", *Journal of the American Society for Information Science*, vol. 48, no. 8, p. 707-717, 1997.

[SRI 04] SRINIVASAN P., "Text mining: Generating hypotheses from MEDLINE", *Journal of the American Society for Information Science*, vol. 55, no. 5, p. 396-413, 2004.

[STE 03] STEGMANN J., GROHMANN G., "Hypothesis generation guide by co-word clustering", *Scientometrics*, vol. 56, no. 1, p. 111-135, 2003.

[SWA 86] SWANSON D.R., "Fish oil, Raynaud's syndrome, and undiscovered public knowledge", *Perspectives in Biology and Medicine*, vol. 30, no. 1, p. 7-18, autumn 1986.

[SWA 87] SWANSON D.R., "Two medical literatures that are logically but not bibliographically connected", *Journal of the American Society for Information Science*, vol. 38, no. 4, p. 228-233, July 1987.

[SWA 88] SWANSON D.R., "Migraine and magnesium: Eleven neglected connections", *Perspectives in Biology and Medicine*, vol. 31, no. 4, p. 526-557, summer 1988.

[SWA 89] SWANSON D.R., "Online search for logically-related noninteractive medical literatures: A systematic trial-and-error strategy", *Journal of the American Society for Information Science*, vol. 40, no. 5, p. 356-358, September 1989.

[SWA 93] SWANSON D.R., "Intervening in the life cycles of scientific knowledge", *Library Trends*, vol. 41, no. 4, p. 606-631, spring 1993.

[SWA 99] SWANSON D.R., SMALHEISER N.R., "Implicit text linkages between Medline records: Using Arrowsmith as an aid to scientific discovery", *Library Trends*, vol. 48, no. 1, p. 48-59, 1999.

[SWA 01] SWANSON D.R., "ASIST Award of merit acceptance speech: On fragmentation of knowledge, the connection explosion, and assembling other people's ideas", *Bulletin of the American Society for Information Science and Technology*, vol. 27, no. 3, p. 12-14, February/March 2001.

[SWA 08] SWANSON D.R., "Literature-based discovery? The very idea", *Literature-based Discovery*, Springer-Verlag, Berlin, Heidelberg, 2008.

[SWI 05] SWINERS J.L., "La sérendipité ou l'exploitation créative de l'imprévu", *Automates Intelligents*, 2 April 2005. Available online at http://www.automatesintelligents.com/echanges/2005/avr/serendipite.html, consulted on 18 March 2009.

[TRE 07] TREMEL N., PIERRET J.D., DOLFI F., LOESCHE C., Administration of tropisetron for treating inflammatory skin diseases/disorders, international patent no. WO2007099069, 2007.

[WEE 00] WEEBER M., KLEIN H., ARONSON A.R., MORK J.G., DE JONG-VAN DEN BERG L.T.W., VOS R., "Text-based discovery in biomedicine: The architecture of the DAD-system", *Proceedings of the AMIA Symposium*, p. 903-907, 2000.

[WEE 01] WEEBER M., KLEIN H., DE JONG-VAN DEN BERG L.T.W., VOS R., "Using concepts in literature-based discovery: Simulating Swanson's Raynaud-fish oil and migraine-magnesium discoveries", *Journal of the American Society for Information Science and Technology*, vol. 52, no. 7, p. 548-557, 2001.

[WEE 03] WEEBER M., VOS R., KLEIN H., DE JONG-VAN DEN BERG L.T.W., ARONSON A.R., MOLEMA G., "Generating hypotheses by discovering implicit associations in the literature: A case report of a search for new potential therapeutic uses for thalidomide", *Journal of the American Medical Informatics Association*, vol. 10, no. 3, p. 252-259, May/June 2003.

[WRE 04] WREN J.D., BEKEREDJIAN R., STEWART J.A., SHOHET R.V., GERNER H.R., "Knowledge discovery by automated identification and ranking of implicit relationships", *Bioinformatics*, vol. 20, no. 3, p. 389-398, February 2004.

[YET 06] YETISGEN-YILDIZ M., PRATT W., "Using statistical and knowledge-based approaches for literature-based discovery", *Journal of Biomedical Informatics*, vol. 39, no. 6, p. 600-611, December 2006.

[YET 08] YETISGEN-YILDIZ M., PRATT W., "Evaluation of literature-based discovery systems", *Literature-based Discovery*, Springer-Verlag, Berlin, Heidelberg, 2008.

Chapter 11

Processing Business News for Detecting Firms' Global Networking Strategies

11.1. Introduction

Competitive Intelligence (CI) has now evolved into an internationally recognized discipline, which is indispensable for the organization's strategy and economic performance [BOT 04]. With increased global competition, sustained and often radical innovations, volatile markets, and strong pressure from shareholders on financial performance, the economic environment of organizations is very difficult to apprehend. In addition to defensive approaches, offensive approaches which advocate a dynamic CI directly linked to the company's decision-making process and geared toward understanding the complex environment in which the company evolves, have become essential.

Research has long emphasized the importance of engaging in competitive intelligence activity and of understanding the ever-changing technical capabilities of any given industry [BRO 98, GLU 94]. This is of paramount importance today, since more unexpected competitors are emerging from "outsider" industries. Companies that use action-oriented CI and possess advanced tools to assess their environment and competitors' capabilities have been shown to outperform companies that do not invest in the intelligence process [SUB 98]. Since timely information gathering is the basis and key of

Chapter written by Brigitte GAY.

CI, knowing how to collect CI in the environment of Web 2.0 is essential as Web 2.0 harnesses collective intelligence.

Web 2.0 refers to the convergence between increasingly more sophisticated tools and their use, be it wikis, blogs, tweets, social software, etc. These new tools enable the emergence of individual, as well as collective behavior. If social networks indeed encourage collaborative activities, any single player (man, company) can also present himself/itself as well as his/its network of correspondents and practices of interest. Owing to the architecture of the World Wide Web, even individual actions can build collective value as an automatic by-product. Tools come with a lot of features such as directories, advanced search operators, classifications by themes and/or interest groups, the possibility of creating links between groups, etc. Web 2.0 promotes customization of practices and social uses of the Internet including information retrieval with regard to tools, data access or their structuring.

The value of these tools and practices is seen in the fact that most of them (blogs, podcasts, online videos, wikis, social networks, etc.) are now being used by enterprises in many ways. Figure 11.1 shows an example of the possible integrated use of social media by enterprises.

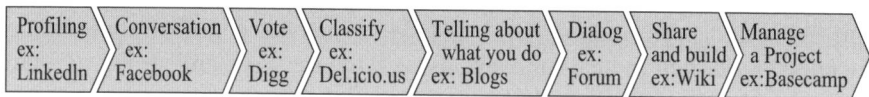

| Profiling ex: LinkedIn | Conversation ex: Facebook | Vote ex: Digg | Classify ex: Del.icio.us | Telling about what you do ex: Blogs | Dialog ex: Forum | Share and build ex:Wiki | Manage a Project ex:Basecamp |

Figure 11.1. *Mass communication tools for use inside and outside companies (source: PriceWaterhouseCoopers, ISPIM'2008 Conference)*

Companies' behavior with regard to these media differs. A statistical and longitudinal study conducted in 2008 in the U.S. by the University of Massachusetts (center for marketing research) has shown that the Inc. 500 (firms having the fastest growth) and the Fortune 500 firms (the more traditional 500 richest companies) have different rates of adoption of these media: the first group adopts them more quickly (e.g. 59% of Inc. 500 had blogs against 15.2% for the Fortune 500, representing 20% growth in a year for the first group against 3.6% for the latter). All companies however proclaimed to make these tools an essential element of their strategy, particularly regarding the use of social networks for marketing practices.

More generally, Web 2.0 is profoundly changing the industrial world in which we operate. Business models, the way to compete, and temporal dimensions are being transformed. Internet users or indirect competitors, for example, can quickly offer online freeware advantageously competing with payware which takes longer time to develop within companies and to be introduced into the market (e.g. Writely against MS Word, Google Spreadsheets against MS Excel). Another well-known example of change is the introduction of an innovative business model – Adwords – by Google, and the pay-per-click concept.

From Web 2.0, we must retain the fact that "social networks" are strong elements of individual and enterprise environment. Networks which are evolving as an output of the collective activity of some Web users can be easily and directly observed on the Web, while data management is a core competency needed to reveal other, hidden, link structures. Web mapping and social network analysis are considered essential for CI in the United States and in Asia.

We will show more specifically the importance of analyzing through Web 2.0 the networks of interfirm transactions that are made and unmade quickly around the globe and which currently constitute a key component of business strategy – as firms resort increasingly to open business models.

11.2. A strong trend: Webs of transactions

Industries are now globalized; competition is very strong and depends on constant global trade of technology and innovative products. The term "global innovation arms race" has been used to emphasize the competitive pressure which forces companies to accelerate their pace of innovation [BAU 02]. The growing number of interfirm transactions has been one of the major responses to this pressure. At present, alliances are continuously being formed worldwide. This movement has resulted in the formation of complex global networks of contracts and has redefined organizational boundaries. New terms have emerged – networked firm, virtual enterprise, open business model – which emphasize the increased reliance of firms on contracts as part of their business models, firms using internal and external paths to market as they seek to advance their technology.

Competition between firms has given place to a new paradigm – confrontation through "portfolios" of alliances – firm performance leveraging on that of other firms in rather ephemeral relationships, as firms constantly redraw their boundaries [HIT 98, HOL 09, NEW 96]. Some authors describe *hubs*, or firms that rely heavily on transactions, as the bright illustration of the ability of companies to change the rules of the game through creative and inventive thinking [LOR 95].

From a more pragmatic point of view, analysts such as Booz-Allen & Hamilton estimate that more than one third of the income of the first 2000 Western companies comes from alliances. Reuer and Ragozzino [REU 06] have observed that for half of the "big" companies, 6 to 15% of the company's value was created from their transactions (30 links or more). Very large companies like IBM and Pfizer currently manage hundreds of alliances. IBM licensing revenues totaled more than US$ 1.9 billion in 2001, up from US$ 30 million in 1990 [CHE 03], while Pfizer studied in 2005 more than 400 contract opportunities, concerning either licensing agreements or acquisitions.

CSC (Computer Sciences Corporation) have reported that in-licensing revenue of the top 20 large pharmaceutical companies reached $63 billion in 2004, up from $38 billion in 2001 and estimated that, by 2010, one full quarter of large pharmaceutical sales would be derived through the licensing channel.

As all knowledge required for product development is becoming highly complex and depends on skill sets increasingly originating from diverse sectors or industries, or a combination thereof[1], networks become loci of innovation which provide quick access to knowledge or resources that are not present in the organization and may take a very long time to develop within its boundaries [POW 96].

Procter & Gamble claim that today 35% of their company innovations, as well as billions of dollars in income, are the result of their open innovation

---

1 For example, gene sequencing constitutes a sector of huge economic interest for the pharmaceutical industry; today, it leverages on nanotechnologies, the ability to process and analyze terabytes of data, and biotechnologies. Different kinds of companies which compete in this sector belong to various different industries, and range from small-size high-tech companies (introducing multiple breakthroughs and rapid changes in the sequencing sector as well as significant cost reductions) to mature biotechnology companies, as well as large pharmaceutical companies. The pace of innovation is extremely strong, and competitive advantages are of short duration. This sector and the biotechnology industry as a whole are defined by a succession of rapid technological phases called next – next – next generations; these terms refer to a volatile market led by successive discoveries brought in by newcomers or players originating from different industries.

strategy [HUS 06]. If joint ventures often epitomized in the past the corporate relationships in a then rather stable universe, what prevails today is the issuing of licensing agreements.

Licensing agreements, with their inherent flexibility, allow companies to respond quickly to continually changing markets. These contracts, exclusive or not, can indeed be adjusted to the specific needs of enterprises, including to given steps in the value chain or geographical areas, and have specified time frames; companies can also turn to *in*[2]- or *out-licensing*[3].

Company's performance is more and more decided outside its boundaries. The managers' choices have become complex: the speed at which they will capture innovation from the outside and the nature of their choices will necessarily have distinct effects on company performance.

Entrepreneurs must thus understand how to build and manage temporary alliances configurations or networks that must enable them to control, and possibly constrain, the highly competitive environment in which their company is situated. The knowledge of the way the alliance networks are dynamically built and the appropriateness of their design would provide them with competitive advantage.

The consequences for CI of the increased blurring of corporate boundaries and of a transaction-based intense global competition are numerous. Dynamic multilevel analysis can be carried out via data aggregation; we can thus analyze the evolution of complex global transaction networks of industries or sectors, but also the adequacy of the position and thus alliance portfolio of any individual firm within complex these entwined environments, and in light of the worldwide rapid technological and business evolution.

Real-time assessment of a company, its partners, its competitors and environment capabilities, and how these different levels are entangled and "fit" together, are possible. If networks of interfirm alliances, like Web 2.0 social networking, are great means of (corporate) renewal and fast-paced action, they also disclose a firm strategy to CI analysts as the firm (re)draws its "boundaries". Similarly, an individual who uses Web 2.0, professionally or not, constantly provides information to CI professionals.

---

2 Agreement whereby a company acquires the right to use some of the assets of another company.

3 Whoever has assets has the right to licence them to another party.

11.3. Leveraging Web 2.0 for analysis of global interfirm trade

Since a well-known example of Web 2.0 applications is Wikipedia, it seems natural to stress key points of the definition of competitive intelligence in this online encyclopedia. It is stated there that "The focus is on the external business environment... The term CI is often viewed as synonymous with competitor analysis, but competitive intelligence is more than analyzing competitors – it is about making the organization more competitive relative to its entire environment and stakeholders: customers, competitors, distributors, technologies, macro-economic data, etc".

What is yet missing from most of the definitions of CI are the concepts of time and connectedness. Environments are unstable, prone to rapid change, agents are interconnected, and open business models allow firms to function very quickly this has been discussed earlier, but we shall give examples later.

One of the consequences of the deep instability of transactional activities and markets is the increased attention which the firm must pay not only to its competitors but also to its partners, and especially how these partners interact with other firms, the rapidly obsolescencing of their core assets, and the (lack of) relevance of their global positioning at all times, etc. Another issue is that any given firm is constrained by constant moves of all other firms in a system. Therefore, an organization, for which anticipating the future was before a main and complex goal, is now also faced with the difficulty of tracking and "reacting" to the constant large-scale changes in its environment.

New strategic imperatives thus emerge. They include continuous multilevel monitoring of industries or sectors to which the firm belongs and being able to anticipate as well as make immediately adjustments to one's strategy. In addition, technical breakthroughs can often come today from very different types of industries and sectors. Indeed, crossovers between very different sectors and industries are becoming the norm rather than the exception (for example the bio-energy industry links white and green biotech sectors with the automotive industry). Modern CI must therefore often transcend industry bounds.

Positioning a firm relative to its entire environment is essential in strategy. One of the maxims of CI is that 90% of the critical information that a company needs is already public or can be systematically developed from public data. Three main steps are required: the collection and storage of data, their analysis and interpretation, and their appropriate dissemination.

The primary sources of information available via Web 2.0 are numerous e.g. e-newsletters, online newspapers and online news such as Reuters, Google news (with possible use of RSS or really simple syndication feeds), Yahoo, Wall Street Journal and NEWSiness alerts, free or paid "business" sites such as Hoover's, LIVEDEDGAR, LexisNexis, Factiva, Dun & Bradstreet, SEC Filings, Business wire, CNN Money, MSN Money Central, Reuters stock information etc. and their aggregators (e.g. PR Newswire), databases (e.g. WIPO, EspaceNet, USPTO for patents; Mergerstat M&A for transactions), videos (e.g. CBSnews.com, YouTube), stock quotes, Web search engines, blogs (e.g. Blogdigger, Technorati) and forums (Omgili.com), plus online tools for monitoring competitors, financial or legal blogs, tools to track changes in company Websites, etc. Company' Websites are also giveaway sources of information about their finances, history, employees, technologies, contracts and patents; photos and Web conferences, updates of companies' sites, and the (change in) language used also contain very revealing elements. Other sources of information include professional social networks like LinkedIn (100 million members) and Viadeo.

A few extremely interesting tools such as Silobreaker and Marketwatch also try to look at the connections between people, companies, places, keywords, topics, etc. in a more or less specific way (Figure 11.2). These tools are not yet fully functional (but should become so very soon) and do not allow processing of evolving data – a key element to understand the dynamic strategic positioning of companies and for technological audits.

Web 2.0 thus verifies the very assumption of CI that information is mostly public. Web 2.0 has indeed made it plethoric and masses of new information are produced continuously. Three interrelated problems which involve very different areas of expertise, however, arise: the need to extract and process very large numbers of multi-source data, the fact that these data are not or poorly formatted for analysis, to the way to unravel the complexity and dynamics of systems to be studied.

Terms related to different efforts made in trying to solve these problems exist such as data and text mining, data parsing, linguistic engineering, cartography for data visualization, complex networks modeling, statistical physics etc. The ultimate goal is to allow users to "continuously" query Web 2.0 in an easy way from anywhere (either from a mobile phone or portable computer) and to immediately obtain query results without having to undergo the computer, linguistic and mathematical processing which underlies elaborate responses.

Among the commercial tools which combine text mining and semantic analysis, we find for example ClearForest TextAnalytics and Goldfire InnovatorTM. An assessment of the strengths and weaknesses of some of the text mining and visualization tools has recently been performed by Bristol Myers Squibb pharmaceutical company [YAN 08].

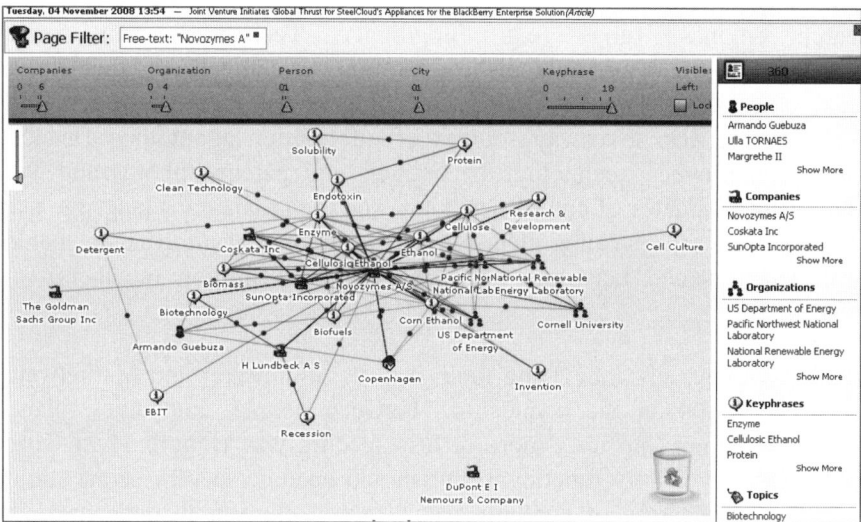

Figure 11.2. *An example of using a media-monitoring product, Silobreaker. A query was formulated regarding the company Novozyme. A network was automatically generated from Web news with the company's name now at its center and including the names of other companies, people, government and other organizations, keywords, and themes, and links between them and with the concerned company. Cursors enable the users to view all the information or part of it. Tracking this information would, in principle, allow identifying and monitoring of key elements of a company's strategy as they appear on the net. This type of software is very promising, and its goals are far more ambitious than the simple example used here*

However, most of the steps toward strategic planning and decision-making based on Web 2.0 are still being treated separately. While some of the steps need to be improved (such as the ability of software to process very large amounts of data quickly, dynamic visualization of networks, and their interactive multidimensional representation), some of them are not yet functional (such as linguistic engineering or mathematical models for complex networks or simulation of "future" networks).

There is rapid progress and a better understanding of the stakes in the U.S. and Asia, where large means are being deployed specifically to solve these problems and in general to fund the information industry.

Examples of analyses that can be made using business news available on the Web 2.0 are given in subsequent chapters to demonstrate the vision that the company can and should constantly have of its environment, partners, competitors, etc. and of itself!

11.4. Companies: "open" but "caught in the Web"

Innovation is now said to be open, since it is exchanged very rapidly through worldwide interfirm transactions. In the pharmaceutical industry, for example, approximately 6,000 contracts were finalized between 2004 and 2005. These global transactions form "Webs" or complex networks that render industrial or sectoral fluxes (technological, process, and product flows). It is now possible to map and analyze all the contractual exchanges which bind the companies. Mapping tools can help a company to evaluate, control, and even impose, its position in constantly evolving financial and economic spheres. "Micro" versus "macro" and dynamic visions of the business environment are essential (Figure 11.3).

Different network layers must thus be considered for analysis:

– large-scale industrial transaction networks;

– transaction networks of industry sectors; all sectors of an industry are analyzed separately. The overall industry image can be reconstructed while taking into account inherent variation in growth rates between sectors;

– transaction networks in a firm neighborhood. They allow clear analysis of a company partners' partners; we can thus capture in real-time the value of the firm partners as perceived by the market as well as their strategic behavior in the face of change;

– the transaction networks created by the company itself, or "egonets", and the relevance and dynamics of these alliance portfolios;

– the firm and how its assets evolve through acquisitions, majority or minority equity participations by or in other companies, exclusive, non-exclusive, or cross-licensing activities, etc.

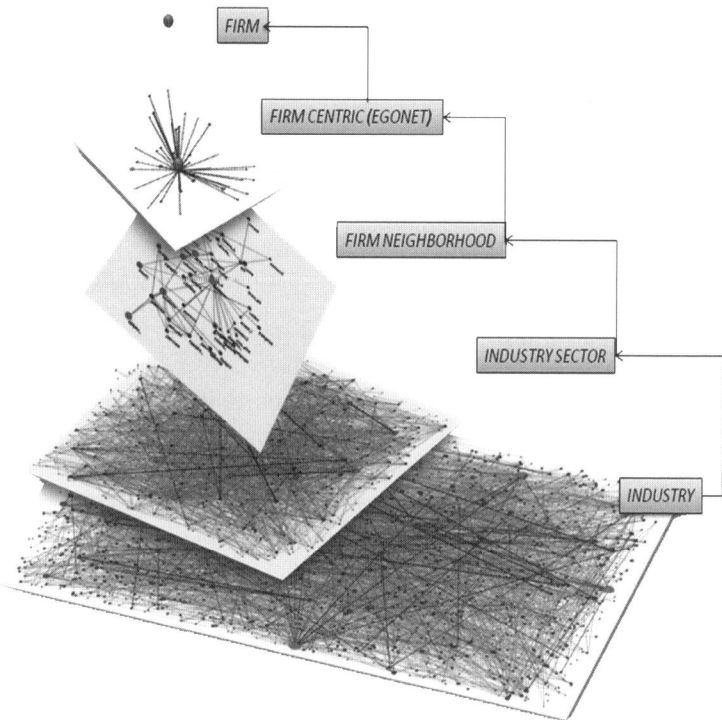

Figure 11.3. *Multilevel analysis of transaction networks*

The following investigations can be carried out:

– sectorial or industrial developments with identification of technological, products, processes, and financial fluxes that circulate throughout networks; the rate of diffusion of innovation in these global and unstable systems; steps of the value chain where contracts are made; detection of dominant (and for how long) and emerging players in each sector of the industry; identifying the different phases in the life cycle of a sector (emerging, growth and, maturity, obsolescence) and their growth rates;

– dynamic benchmarking of competitors and/or partners, their alliance strategies and risk-taking capabilities, the turnover of their contracts, the adequacy of their positions at all times (at national and international levels); determination of the sectors firms have invested in and of how these investments fare compared to that of competitors;

– dynamic trade analysis by country (by city), identification of dominant and emerging countries and their expansion policies: have they invested in key sectors? How are they connected to other countries? Are they deficient regarding technological learning and assets, products, etc.?

– temporal analysis of companies' ties with countries in the world;

– analysis of the financial structure of contracts.

Although it is important to also make use of state-of-the-art methods, disciplines, and practices of systems modeling and simulations, we will only demonstrate below how to visualize open-source data using mapping tools, considering the pharmaceutical industry as an example.

There are many freewares on the Web, but few are able to represent data dynamics. We have used here to showcase our examples the software program VisuGraph, which is still currently in a working prototype form [GAY 09]. Business news briefs on the Web contain a lot of information despite the shortness of their text. We used data parsing to convert functional units into machine language which is more easily managed (Figure 11.4).

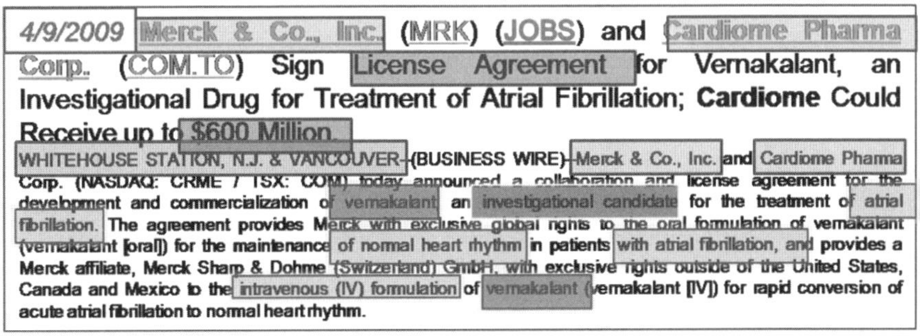

4/9/2009 Merck & Co., Inc. (MRK) (JOBS) and Cardiome Pharma Corp. (COM.TO) Sign License Agreement for Vernakalant, an Investigational Drug for Treatment of Atrial Fibrillation; Cardiome Could Receive up to $600 Million.

WHITEHOUSE STATION, N.J. & VANCOUVER-(BUSINESS WIRE)-Merck & Co., Inc. and Cardiome Pharma Corp. (NASDAQ: CRME / TSX: COM) today announced a collaboration and license agreement for the development and commercialization of vernakalant an investigational candidate for the treatment of atrial fibrillation. The agreement provides Merck with exclusive global rights to the oral formulation of vernakalant (vernakalant [oral]) for the maintenance of normal heart rhythm in patients with atrial fibrillation, and provides a Merck affiliate, Merck Sharp & Dohme (Switzerland) GmbH, with exclusive rights outside of the United States, Canada and Mexico to the intravenous (IV) formulation of vernakalant (vernakalant [IV]) for rapid conversion of acute atrial fibrillation to normal heart rhythm.

Figure 11.4. *Example of business news taken from Business Wire site. Information on the date, the companies that interact with each other, the places, the technologies or products and their name, the indications, the stages of the value chain, the formulations, the amount ($) involved, the legal nature of the contract, are cut into blocks to be processed and cross-analyzed*

11.4.1. *Macro network: industry and market segments*

Figure 11.5 shows the contractual intensity which characterizes most of the industries today. About 5,000 firms have been transacting in the pharmaceutical industry worldwide in just 2 years. The most active companies are the biggest pharmaceutical companies like Pfizer, Novartis, etc.

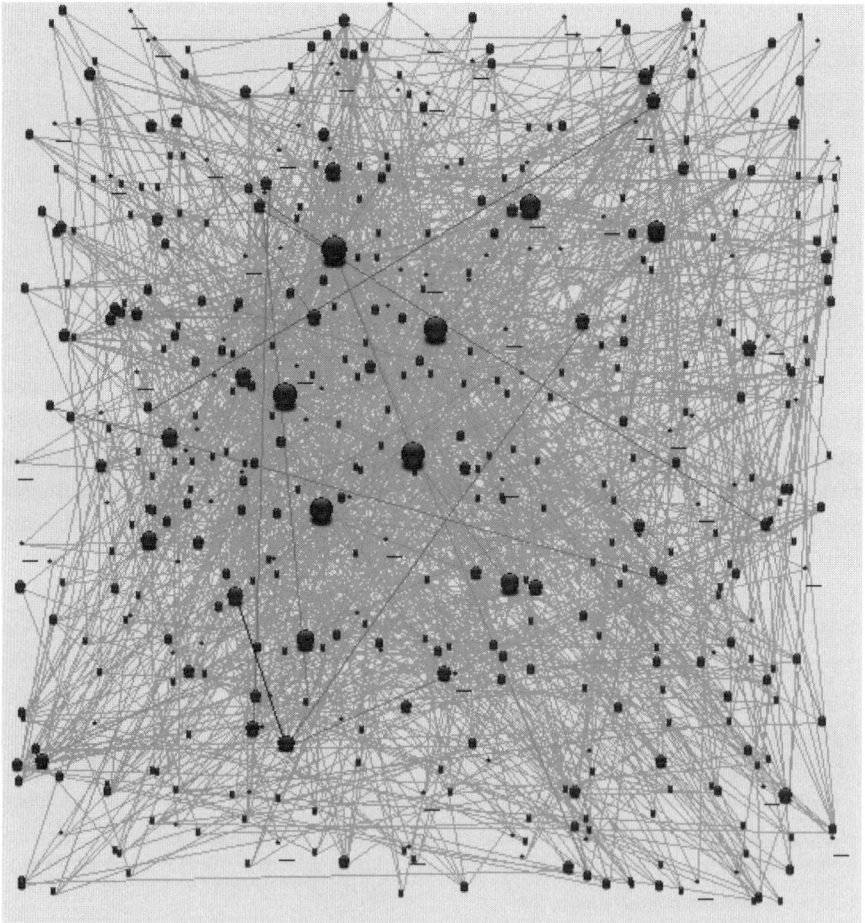

Figure 11.5. *Transaction networks in the pharmaceutical industry between 2004 and 2005. The nodes are firms, their size being proportional to the number of transactions that the company makes, and the lines that tie the nodes represent the contracts*

The size of a company was determined in the past by its ability to manage alone all the steps of a value chain. We can clearly see on the graph that this is no longer the case, and that big companies have open business models.

The global dynamics of complex interfirm networks is not random. Figure 11.6 shows an alliance network in a major sector of the biopharmaceutical industry and compares it with a network in which the same number of players having the same number of links, interact randomly. The real-world network has a non-trivial characteristic "scale free" structure, with few highly-connected firms or *hubs*. These structures are often found in industries.

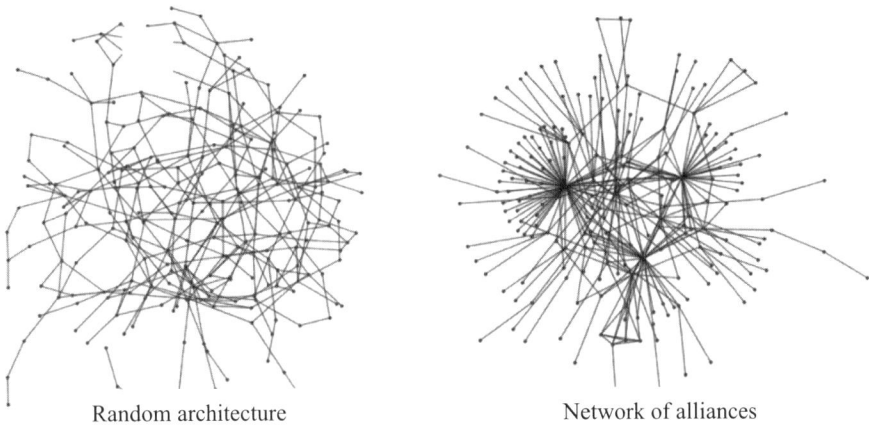

Random architecture   Network of alliances

Figure 11.6. *A network of alliances in a major sector of the pharmaceutical industry over a period of 3 years (right). A random network containing the same number of nodes/firms and links/contracts is shown on the left*

At the industry level hubs are major companies that continuously capture assets outside their boundaries whereas at the sectoral level hubs are young and highly innovative companies specialized in a given sector, and which sell and distribute their unique assets in this one sector.

Changes in the architecture of networks reveal changes in the power structure of an industry or a sector. Figure 11.7 shows the dynamics of a major sector of the pharmaceutical industry. This sector is highly unstable, as it is subjected to a very high turn-over of firms (75% of newcomers each year).

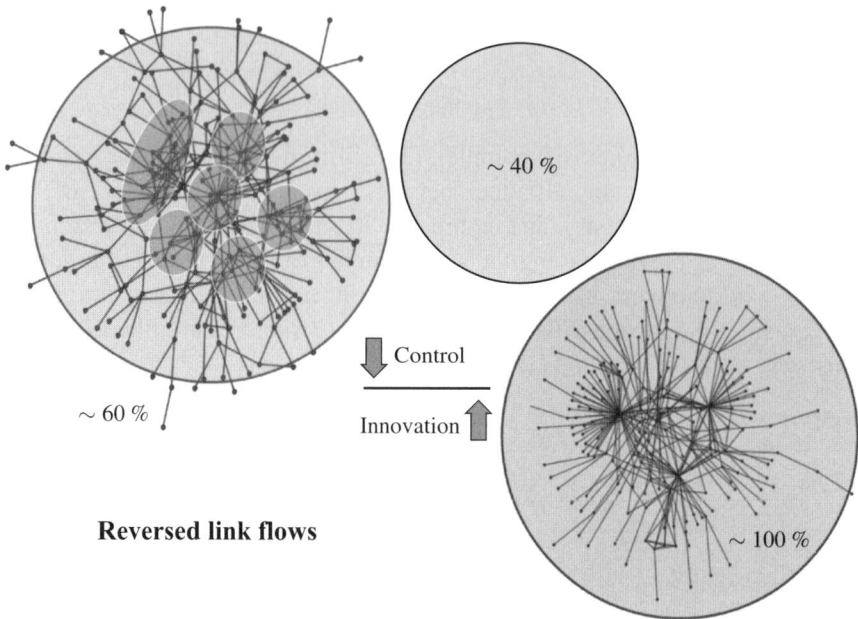

Figure 11.7. *The emerging sector (2000-2002; bottom right) is dependent on three hubs which diffuse (mostly out-license) their breakthrough technological assets (or products derived from these assets) across the network. The system is unstable, with the continuous arrival of newcomers resulting in the rapid disappearance of preceding innovative players. In the last plot (upper left, period 2006-2007) innovations are now only incremental and brought in by many new entrants (detailed analysis of the business news). If in 2000-2002 a few breakthrough innovations were disseminated throughout the graph, in 2006-2007, cohesive communities (dark gray circles) are clearly apparent. These cohesive pockets are controlled by major pharmaceutical companies which are essentially in-licensing technologies/products. All transactions were contained in a single network in the first period; in the last period, however, the main network is representative of only 60% of transactions; 40% of the transacting activity has formed outside this network, thus heralding a new technological cycle*

11.4.2. *Companies and portfolios of alliances*

In the pharmaceutical industry, major players remain rich (e.g. they are top-performing companies from 2004-2009 as ranked by Forbes 2000); it is, however, interesting to note that they use a high turnover of transactions to cope with global change and uncertainty (Figure 11.8).

Figure 11.9 shows that we can capture in a few clicks, the companies' ability to manage change and risk taking in any sector or industry.

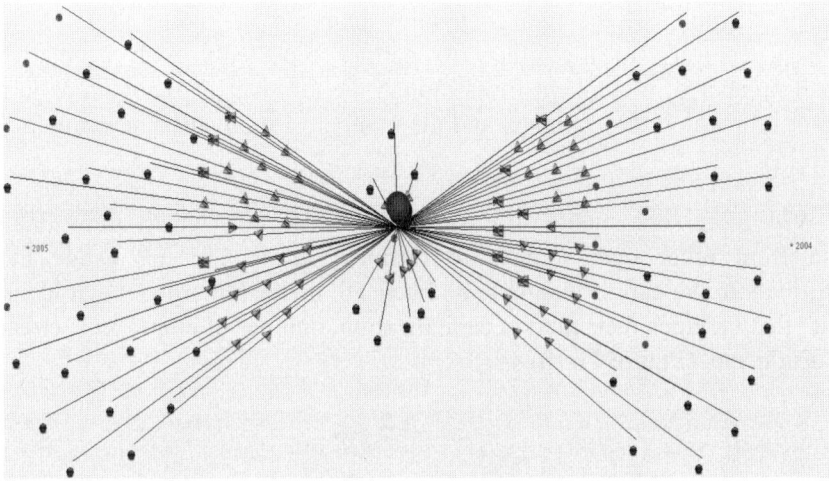

Figure 11.8. *Transactions turnover of a big pharmaceutical company from 2004 (right) to 2005 (left) in the entire industry. Each year the company (at center of the image) makes many new transactions with many new partners. The eight partners with which it interacts repeatedly are located in its vicinity. The links are arrowed and show mainly in-licensing activities. In 2004, the company disinvested several assets (links to the black triangles)*

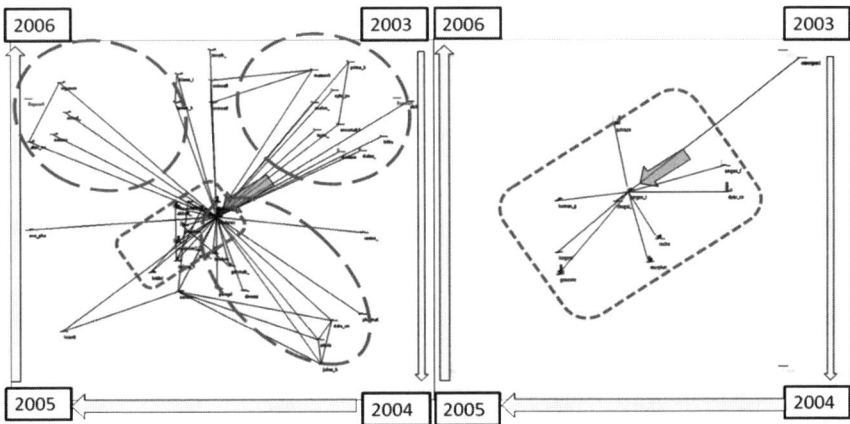

Figure 11.9. *Analysis of the turnover of contracts of two companies (blue arrows) over 4 years (2003-2006) in a major sector of the biopharmaceutical industry. Nodes or companies involved in 1 year are placed at the periphery of a "dial" and close to their temporal marker or year. Those involved in more than 1 year are positioned closer to the center of the dial and between their temporal markers or years of presence. The two graphs show that the company on the left is constantly interacting with newcomers, while the company on the right relies on a few, incumbent, players*

For example, of the two companies analyzed in Figure 11.9, one has chosen, contrary to the other, to partner repeatedly with a relatively small number of incumbent firms though the turnover of innovations in the sector into which they operate, and the number of innovative newcomers are high.

With more sophisticated use of the analytical tools which allow linking data from other open-sources to business news, we can gauge more information: we can for example add beneath nodes (enterprises) an identity "kit" that includes their market capitalization, number of employees, country of origin, etc. (Figure 11.10).

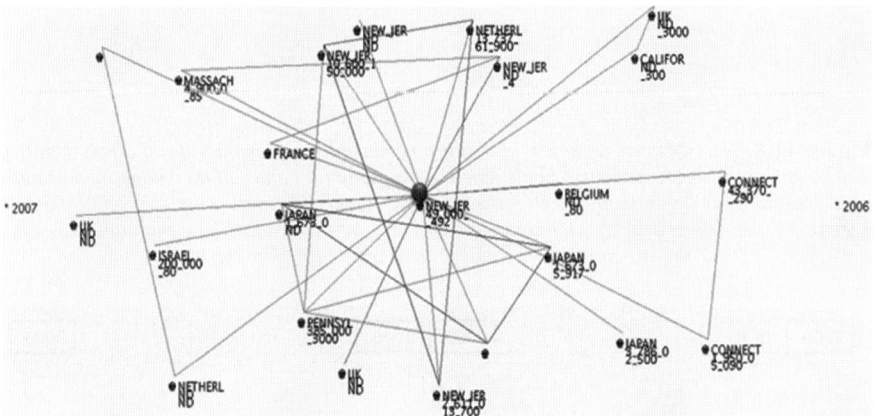

Figure 11.10. *The alliance network of a company (center) is extracted following the data parsing of all business news in a representative industry. One can see the alliances the firm makes from 1 year to another (2006 and 2007; right to left). Identity kits (here country, market capitalization, number of employees) obtained from Web sources such as Yahoo! Finances allow a better understanding of the type of partners the firm interacts with. We can see very quickly that this firm in New Jersey has global contracts, even though four of its partners are also from New Jersey*

11.4.3. *The dynamics of global competition*

We have only exploited some of the data present in business news (mainly dates and names of companies). The idea was just to demonstrate rapidly that management of data on Web 2.0, including complex network analysis, taking interfirm alliance networks as an example, was crucial.

In this final section, we will show that one can rapidly have spatial data, and we will again demonstrate the importance of dynamic studies. Of course, the simplest method is to generate bar charts or maps of continents, countries, regions, cities, where the transactional activity is greatest. A map is shown in Figure 11.11.

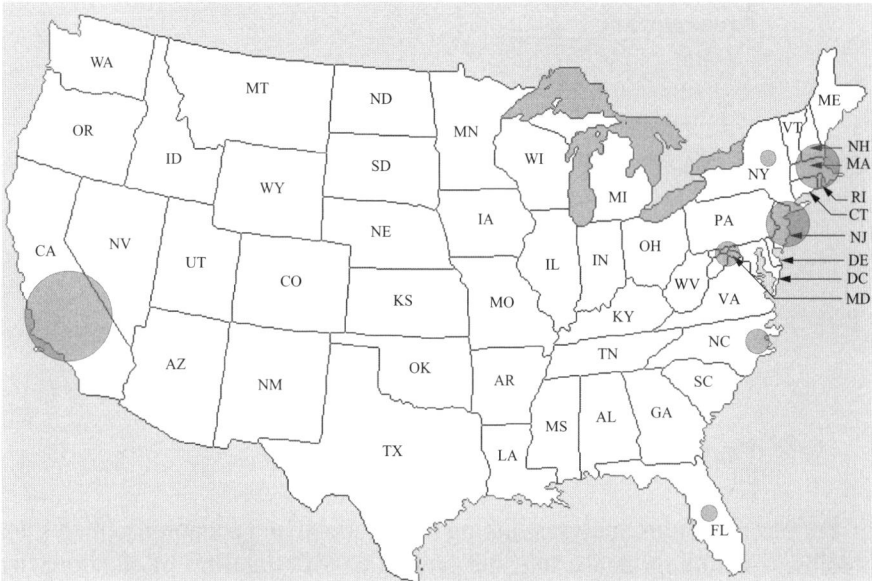

Figure 11.11. *Transactional activity of the various U.S. states in the biopharmaceutical industry (2004-2007). The size of the circles is proportional to the level of activity. Major areas are the East Coast and California*

More sophisticated analyses involve considering the exchange dynamics between country/regions/cities. For example, one can observe the rise of Asia in the pharmaceutical industry in more recent periods though Western countries still generally dominate the industry (Figure 11.12). More detailed studies show that China was very dependent on the West in 2004 and Asia was not developed and that, within a year, Asia had grown rapidly. China is at the center of this growth, poised to become the leading country in the Asia pharmaceutical space, and getting off its dependence on the West (Figure 11.13). Advanced online language translation tools will play a crucial role in helping to monitor ongoing developments in this area and elsewhere.

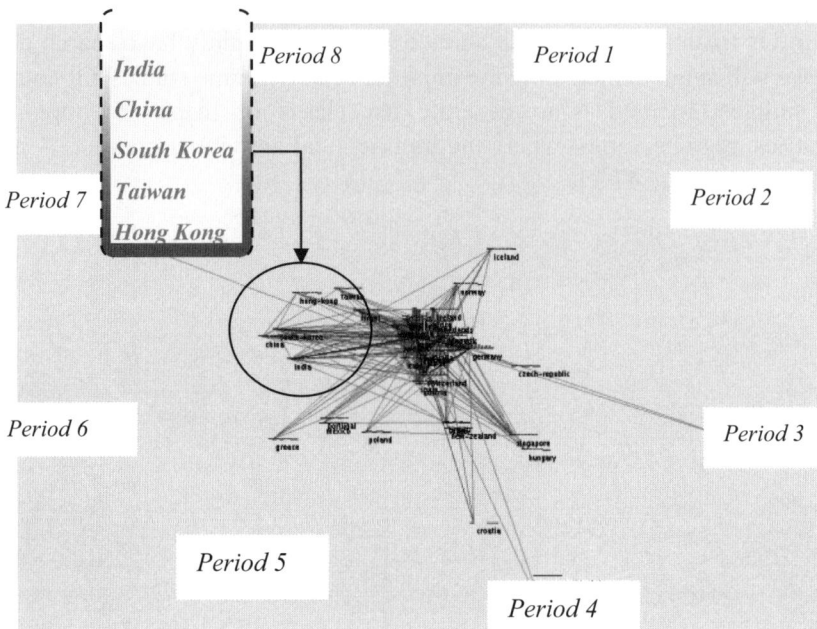

Figure 11.12. *Transactional activity between countries over 8 years*

Further numerous analyses linking the financial and economic spheres are possible, including analyzing the volatility or stability of funding by institutional actors in different regions/countries, the origin of these actors, external or belonging to the region/country, etc. The world of Web 2.0 analysis is infinite. Results also show the danger of averaging results, as currently done.

11.5. Conclusion

We have witnessed in recent years the rise of the information industry, the extremely rapid evolution of digital tools and their potential. Processing and analyzing terabytes of data is now feasible. With Web 2.0, there is a new game afoot. Those who understand how to play the game can have a direct impact on organizations, their functioning, and "connectedness".

Moreover, globalization, the subprime crisis, competitive pressure for innovation, "low cost" strategies, and speed of trade are destructuring industries and force firms to redefine themselves; they now have fuzzy boundaries and constantly change their strategies.

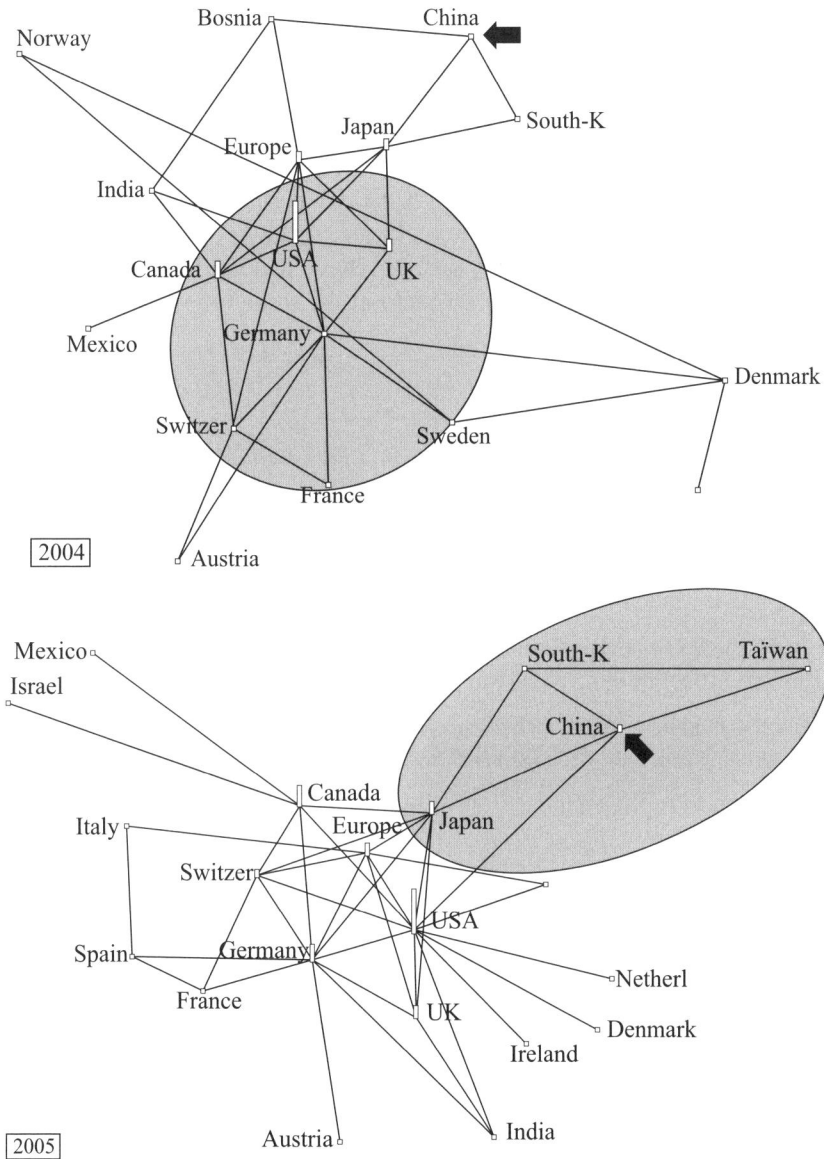

Figure 11.13. *Global interfirm transactions (see Figure 11.5) resulting in intercountry collaborations in 2004 (top) and 2005 (bottom). The gray circles show the beginning of Asian development, with China at its center, in 2005, in contrast to the observed dependence of China on the West just a year before. We also see on this graph that in 2005, although the UK and Germany have about the same number of transactions, the position of Germany is better. Other findings are the extension of the United States to new European markets, markets that the major European players do not seek, and the weak connection of European countries with Asia*

All this forces us to reconsider CI and corporate strategy. Network analysis is one of the new points through which tangible progress is possible. We have shown that many networks are detectable on Web 2.0. Though forming networks is an essential element of corporate strategy, open business models also result in exposing firms to manipulation and coercion. An offensive, "virtual" and network-based, CI is indispensable for a firm's strategy and should be conducted in real time, so as to give corporate leaders the means to strongly and durably position their firm worldwide… Networks can constitute major assets and drawbacks at the same time.

11.6. Bibliography

[BAU 02] BAUMOL W.J., *The Free-Market Innovation Machine. Analyzing the Growth Miracle of Capitalism*, Princeton University Press, Princeton, Oxford, 2002.

[BOT 04] BOTBOL M., VERDIER I., *France: Le Top 100 de l'intelligence économique*, Indigo, Paris, 2004.

[BRO 98] BROCKHOFF K., "Technology management as part of strategic planning – some empirical results", *R&D Management*, vol. 28, no. 3, p. 129-138, 1998.

[CHE 03] CHESBROUGH H., "The logic of open innovation: Managing intellectual property", *California Management Review*, vol. 45, no. 3, p. 33-58, 2003.

[GAY 09] GAY B., LOUBIER E., "Dynamics and evolution patterns of business networks", *ASONAM Conference*, Greece, 2009.

[GLU 94] GLUECK W.F., JAUCH L.R., *Business Policy and Strategic Management*, McGraw Hill, New York, 1994.

[HIT 98] HITT M.A., KEATS B.W., DE MARIE S.M., "Navigating in the new competitive landscape: Building strategic flexibility and competitive advantage in the 21st Century", *Academy of Management Executive*, vol. 12, no. 4, p. 22-43, 1998.

[HOL 09] HOLMBERG S.R., CUMMINGS J.L., "Building successful strategic alliances: A strategic process and analytical tool for selecting alliance partners", *Long Range Planning*, vol. 42, no. 2, p. 164-193, 2009.

[HUS 06] HUSTON L., SAKKAB N., "Connect and develop", *Harvard Business Review*, vol. 84, no. 3, p. 58-66, 2006.

[LOR 95] LORENZONI G., BADEN-FULLER C., "Creating a strategic center to manage a web of partners", *California Management Review*, vol. 37, no. 3, p. 146-163, 1995.

[NEW 96] NEWMAN V., CHAHARBAGHI K., "Strategic alliances in fast-moving markets", *Long Range Planning*, vol. 29, no. 6, p. 850-856, 1996.

[POW 96] POWELL W.W., KOPUT K., SMITH-DOERR L., "Interorganizational collaboration and the locus of innovation: Networks of learning in biotechnology", *Administrative Science Quarterly*, vol. 41, p. 116-145, 1996.

[REU 06] REUER J.J., RAGOZZINO R., "Agency hazards and alliance portfolios", *Strategic Management Journal*, vol. 27, p. 27-43, 2006.

[SUB 98] SUBRAMANIAN R., ISHAK S.T., "Computer analysis practices of US companies: An empirical investigation", *Management International Review*, vol. 38, no. 1, p. 7-23, 1998.

[YAN 08] YANG Y.Y., AKERS L., KLOSE T., BARCELON YANG C., "Text mining and visualization tools – impressions of emerging capabilities", *World Patent Information*, vol. 30, p. 280-293, 2008.

Chapter 12

Information Property and Liability in the 2.0?

12.1. Introduction

Web 2.0 reflects a shift from interactivity to interaction [LUC 09] and thus contributes to the building of networks that are no longer based on information sharing, but rather on knowledge sharing. This is a connectivity revolution because two effects can be recognized in this networking: first, the reality of interconnection of information through metadata and interoperability of the language used (XML) and second, a network of actors who are motivated to develop social relationships and editorial activities. The values of these fields are: collaboration, innovation, and lack of hierarchy in an environment that is characterized by an abundance of information in a flat world. Therefore, it is not only a technical phenomenon but also a change in our mode of functioning, organizing, learning, and decision-making [QUO 08] in all sectors of the society and is designated by the 2.0 concept. The phenomenon answers to "a desire for a society where trade would be more horizontal, reticular or network, offsetting utilitarianism development" [MAI 08]. The vertical hierarchy no longer exists, and speeches and organization types change, so that personal involvement and ephemeral teamwork around interdisciplinary projects are enhanced. Versatility and flexibility are the watchwords. Innovation and, generally, scientific, artistic, and industrial "creations" are becoming more complex. The 2.0 concept refers to an anthropological paradigm shift. In fact,

Chapter written by Arnaud Lucien.

communication models have evolved and are now realized in a peer-to-peer or many-to-many relationship. Thus, in addition to tools and applications from Web 2.0, all areas of social activity are affected. At present, people use expressions like Enterprise 2.0, Management 2.0, Education 2.0, and this applies as well to the title of this book: Competitive Intelligence 2.0. The professional world is working in networks within a framework of multilingual community. In scientific research, many contributions, sometimes tiny but critical, characterize innovations. We can no more identify a single "inventor", since paternity is shared between a multitude of contributors. This 2.0 spirit has an impact on traditional industrial sectors. It also manifests in non-profit sectors with examples like free software and online publishing activity through open encyclopaedias such as Wikipedia and citizen journalism. This leads to new business and publishing models and new issues related to ownership, protection, and accountability of information.

Economic and publishing models 2.0 raise issues concerning the place of the user who is called to participate. They rely, particularly on the creation of an audience and on the free nature of Web 2.0. The many-to-many communication relationship is expressed by the possibility of publishing information, producing, selling, and buying simultaneously by freeing ourself from traditional distribution models[1]. New forms of publishing have emerged, especially with blogs and wikis. Wikis are social Websites running on the basis of the participation of contributors who have the ability to change and develop pages. The most widely used is Wikipedia.

The main resources of "economy 2.0" are rooted in advertising and are related to the audience generated by the applications. These have impacted the culture and media industry. Indeed, users are highly encouraged, particularly by the most active financial interest and media coverage of a few success stories to increase publications to ensure an adequate audience, both

---

1 It is necessary to mention the "long tail effect". This expression of Chris Anderson refers to the fact that "products that are subject to low demand or have low sales volume can collectively represent a market share equal to or greater than the best sellers if the distribution channels can offer enough choices and create the link to discover them". The distribution patterns of Amazon, Netfix of Rézolibre, and Wikipedia are examples. According to this, less read articles from Wikipedia have collectively more readers than the main items available on other encyclopedias. This "long tail" effect is felt in all sectors of social life and especially in domains like direct sales, energy production, digital, media, tourism, etc.

for their personal interest (ego) and the economic interests of the operator[2]. According to Gilberto Gil, songwriter, Brazilian Minister of Culture[3] (2003-2008), "today, the role of the intermediary is finished. With peer to peer, mechanical distribution system that has brought so much money for so long to the music industry has become obsolete".

Some of the "bloggers" have managed to build legitimacy through the audience generated by their publications. Most of the applications are free but funded by advertisement. The community aspect of Web 2.0 prompts the user to disclose personal information in order to participate in social networks offered by different applications and for organizing specific commercial profiling. Moreover, through the semantic aspect of Web 2.0, adverts are automatically personalized based on user profile and pages consulted.

Gratuity and advertising thus appear as the guiding principles of Web 2.0 in the race for gaining audience which alone allows the sustainability of the model. The goal then is to reach a broad but individually targeted audience to ensure the profits of the operator. Users are given a prominent place in this process, since their contributions ensure the functioning of the system. The models of creation are impacted by questioning the status of the author. Online publication is freed from the notion of authorship, since anonymous contributors participate in the sustainability of joint projects by investing their knowledge and time in collective intelligence.

The examples are numerous: development of Web applications as widgets, writing or improving free online content (Wikipedia, etc.), file sharing: royalty-free pictures or video, and comments and assessments through "tagging" using metadata (tags).

Therefore, new business and publishing models 2.0 introduce new legal issues that fall within Competitive Intelligence 2.0. The first relates to the intellectual property and the status of the author. Indeed, "contributory"

---

2 Non-professional "authors" based on their talent can emancipate from the obliged demands of production houses to create an audience. Community sites such as MySpace and multiple relays offered by the Internet (WebTV) have allowed young artists, previously unknown, to become famous by avoiding the obstacle characterizing the signing of a production contract.

3 "Condividi Condividi la Conoscenza: la cultura incontra la rete" conference, Thursday 9 June 2005, Venise, Scoletta dei calegheri, Campo San Tomae. Gilberto Gil has been very involved in finding a balance between the common interest and the agenda of the capitalist world (Le Monde, 12.07.2009; El País, 12.07.2009).

creation model involves the definition of a legal framework for challenging the traditional prerogatives of the author of a work. The second is the issue of personal data. Indeed, the hyperconnectivity of 2.0, coupled with an increasingly accurate personal data profiling, is a source of profit with the risk of violating privacy and personal freedoms of the user. Finally the 2.0, and the freedom to communicate, produce, share, and disseminate information raise the issue of questioning the responsibility of contributors and operators of 2.0 applications.

12.2. Information Property 2.0: questioning authors' status

Guarantees related to the principle of due process preclude a too frequent adaptation of the rule of law. The legislature time is often criticized, but ensures respect for republican principles and has established itself as a guarantee of the rule of law[4].

There is a cultural divide between the world of 2.0 and the law. The former obeys the rules of constant questioning (perpetual beta), an immediate adaptation to the contingencies of the technological, social, and even legal environment. The latter is gradually adapting to the "evolution of morality". It is therefore the role of the jurisprudence to interpret the law in order to facilitate its application. In the information and communication technology field, rapid developments both technological and practical are often poorly understood by the rulers.

To realize this, we had noted that most of the Western countries have criminalized illegal downloading of documents without necessarily considering the transition to an economy of flows – the latter modifying the relationship between the user and the document. The changes brought about by 2.0 imply thinking of new models of remuneration for authors.

---

4 The principle of due process that results from the rule of law is devoted to Community law. In France, the State Council, in its annual public report of 2006, defines it thus: "The principle of due process requires that citizens should, without an insurmountable effort, be able to determine what is allowed and what is forbidden by law. To achieve this, the standards laid down must be clear and intelligible, and should not be submitted to too frequent and unpredictable changes". The State Council, Public Report 2006, the French Documentation, Paris, 2006.

12.2.1. *Information property versus 2.0*

Peer-to-peer communication enables a many to many communication; users henceforth have the ability to communicate and share all sorts of information goods. In this context, the concept of intellectual property is misunderstood by the society. In fact, it is based on an intangible property right. The infringement of such rights is not regarded by most users as a theft because the enrichment of those who download illegally does not obviously imply an impoverishment of the author.

But damage appears to be constituted for intermediaries in the music and audiovisual industry. This has been a question of debate for nearly a decade. This sensitive issue was initially treated by the jurisprudence, then the legislature intervened: in France, by formulating the "Copyrights and Related Rights in Information Society" law of August 2006 and "Hadopi" act of May 2009; in the United States, by the Digital Millennium Copyright Act of 1998; in Italy, by the "piracy" act of August 18, 2000 and the Urbani act of May 21, 2004; in Spain, the "global action plan against piracy" of April 2005, etc.

In all cases, governments have faced serious challenges relating to the appropriateness of such devices being considered in the light of changing technologies and uses. In fact, the 2.0 concept fits into an economy of flows for which the mechanical model of author's remuneration is no longer appropriate.

12.2.1.1. *The strength of the traditional model of author's remuneration*

The traditional model of author's remuneration is based on the recognition of intellectual property through copyright protection. The English-speaking and the continental system initially had significant differences in the recognition processes. These were primarily in the apprehension of the moral rights[5] of author, and were later erased by the multiple international harmonization efforts (treaties of the World Intellectual Property, EU directives, etc.). Author's remuneration[6] is thus

---

5 Author's moral rights are inalienable, non-transferable, and perpetual. They give certain powers to the author of a work: the right of authorship, the right of respect for the work, the right of first disclosure, and the right of withdrawal.

6 Author's remuneration is realized in the exercise of his patrimonial rights: these are alienable and prescriptible.

organized for 200 years around two prerogatives: the right of reproduction and the right of representation. The first captures the duplication of works in all media, and the second relates to all communication of a work to an audience. In case of multiple authors, their relations are generally organized by contract. We distinguished: collaboration work in which we can identify the specific contribution of each of the sponsors; composite work, which is an aggregation of works by several authors; and the collective work, rights of which are conferred to a person under whose name it was disclosed. Infringement of copyright is severely repressed by the offense of counterfeiting. In this case, it is this criminalization that has long been used to penalize the illegal downloading of copyrighted works. This is so in the United States with the adoption of the Digital Millennium Copyright Act of 1998, and as well in France, since the censorship of the "Creation and Internet" law by the Constitutional Council in June 2009. This question opened a real social debate beyond partisan divisions.

In France, the parliamentary debates of late 2005 allowed us to believe the outcome of a consensus. Thus, the National Assembly voted in, in exceptional circumstances, the legality of downloading between Internet users and the principle of the statutory license – the process providing a fixed remuneration for authors by the Internet users, which were close to the remuneration for private copying passed by the Parliament in 1985. The initiative, contrary to government projects but adapted to the apprehension of the economy flows, was subsequently censured by the same parliamentary chamber in favor of new penal measures to ban downloading. The legislation, which was finally passed, provides a scale of penalties based on acts committed: downloading, making available to the public, etc. This solution remains elusive since it was confronted for the first time by the censorship of the Constitutional Council[7], which took the opportunity to reiterate the great value placed on property right, even immaterial: the downloading and/or making available to the public under the same offense of counterfeiting. The law introduced a new repression: a specific offense created against the creators and distributors of file-sharing programs. The legislature defined peer-to-peer software as a "device obviously intended for making works or objects protected available to the public without

---

7 See Constitutional Council, decision no. 2006-540 of 27 July 2006, http://www.conseil-constitutionnel.fr/conseil-constitutionnel/francais/les-decisions/depuis-1958/decisions-par-date/2006/2006-540-dc/decision-n-2006-540-dc-du-27-juillet-2006.1011.html, available online on 10 May 2009.

authorization" and acted drastically by imposing a complete ban on production of this type of device[8]. The offense is punished severely: the new Article L335-2-1 code of intellectual property provides a sentence of 3 years imprisonment and a 300,000 Euros fine. Software for "collaborative work, research, or exchange of files or objects not subject to the payment of copyright" is not covered by this prohibition. The legislature refused to censor "Collective intelligence" tools. However, the wording is sufficiently vague, thereby leading to multiple interpretations, and based on the multitude of practices permitted by the existing software, we may ask: where is the distinction between authorized devices and unauthorized devices? Therefore, the French judges condemn most popular sharing tools belonging to Web 2.0 technology: Dailymotion, YouTube, MySpace, etc. The reasoning behind this is based on the concept of "infringement by contribution" offense.

An example can be found in the case of Dailymotion which was sued on a civil basis and held accountable for the provision of "means to make a counterfeit" for uploading on its site the entire "Merry Christmas" film by Christian Carion and was fined 13,000 Euros as damages.

The responsibility of operators of contributive applications is changing: from the status founded by the European directive establishing the framework of host liability and submitting it to *a posteriori* liability, even if

---

8 In the United States, the Supreme Court had already faced the same problem at the time of its "Grokster", decision issued June 27, 2005 and previously in 1984 in the case of "Sony-Betamax" (Sony Corp. of America v. Universal City Studio, 464 U.S. 417 (1984)). On the occasion of this precedent, wherein the legality of the VCR was in question, the Court stated that the liability of the person who offers a product capable of substantial lawful use could be engaged, even though this product would infringe its uses. The software also allows royalty-free peer-to-peer file sharing, so their editors were not to be condemned. It was also the first to be held before judges: legal uses having been evaluated by experts to 10%, it was enough to prove the existence of substantial non-infringing uses. The Supreme Court brought in connection with the case "Grokster" a corrective to this interpretation. If it does not question its jurisprudence "Sony-Betamax", it considered that the existence of legitimate uses does not justify *a priori* the behavior of all those who provide technology to acts of infringement. Thus the liability of the software vendor could be initiated either for inducing infringement, or to have benefited without having taken steps to prevent illegal use of their software.

hosts still see their liability engaged *a priori* mainly in respect of intellectual property rights[9].

By the effects of a repression policy, the traditional model remains well established. However, the Web 2.0 revolution has demonstrated the creativity of a new culture which can break the traditional model of mechanical distribution.

12.2.1.2. *The apprehension of a difficult economy flows*

According to Pierre Veltz[10], "flow of goods, capital, technology, pictures, and cultural references: the global economy is that of mobility, with the notable exception of the less skilled, or people born in the poorest countries. This fluidity is accompanied by a fragmentation of production systems now being deployed across the world, relying on networks of sites becoming highly homogeneous in terms of technological level". We must now consider this context, to think of new models of authors' remuneration. A study of the Next Generation Internet Federation has enabled us to identify avenues for promotion of artistic creation in the digital economy, where the economic value of a recording is increased, by transforming it into a personal "experience". By multiplying its formats and operating "slot", we can enhance the relationship with the artists, and the online "fan club" syndication, and merchandize for the Star Academy, through artist "friends" from MySpace; develop the economic value of concerts through advertising and sponsorship, complementarity with the disk or remote and/or delayed access; promote the construction and enrichment of one's musical universe: live "one's" music where one wants, when one wants, and however one wants, but also share and expand one's musical universe, offering services to facilitate access of the unknown to means of production, dissemination, and exchange with communities of amateurs etc.; develop the use of music as "additional value" associated with other products and services: trademarks, spaces, products, services, etc. and organize direct payment by consumers: purchase (with an infinite

---

9 This is particularly the case with the law of "confidence in the digital economy" of 2004 (Lcen), in France, which only allowed engaging their liability following an ignored formal notification (liability *a posteriori*), regularly in intellectual property, particularly under the rule of law "copyright and related rights in Information Society" of 2006, their liability is *a priori* i.e. they are under an obligation to monitor broadcast content and thus require expensive filtering.
10 Pierre Veltz, *Le Monde*, 29 December 2005, "L'économie mondiale à l'ère des flux", http://www.lemonde.fr/savoirs-et-connaissances/article/2005/12/29/pierre-veltz-l-economie-mondiale-a-l-ere-des-flux_725351_3328.html, available online on 10 May 2009.

variety of payment methods), "location", subscriptions, bundling, subscription, donation, etc. and payment by others: public broadcasting licenses, advertising, major portals, manufacturers of portable, ISPs, etc. and digital rights management systems as sources of measurement and collection of value, rather than "protection" against the consumer".

The report's findings reveal in particular the difficulty of the traditional model to remain standing in the face of the gratuity (free) and the "long tail" effect, to ensure the sustainability of a model. The latter seems to be focused toward an economy of flows, "which is moving from an economy based on unit prices and low quantities to high volumes, and low and even immeasurable unit prices, so the consumer only pays for access right to streams; [a] service economy, which finds the path of rarity, uniqueness, and exclusivity in the musical experience, the relationship with a work or an artist, [an] attention economy, ...intermediation between a superabundant, diverse, global "offer" and an increasingly individualized and mobile demand".

12.2.2. *Alternative licensing and contributive phenomenon*

Web 2.0 is usually called "contributive Web" or Writable Web. This technical evolution has an influence on social relations. "The peer-to-peer phenomenon results in a means of production that can coordinate up to thousands of voluntary contributors for the sole purpose of completing a project without compensation" [QUO 08]. In addition to integrating the works into an industrial circuit and then making them available online and sharing with applications, the 2.0 spirit is at the origin of new types of creation: individual creation with a non-profit goal and collaborative works capable of bringing together hundreds of contributors. The first case corresponds to a "creative" author of videos, photographs, text, etc. who wants to publish his/her works on a sharing site such as WebTV, on online social networks such as the popular MySpace, or an open encyclopaedia like Wikipedia.

With these uses, Creative Commons licenses became very popular and allowed enforcing the moral rights of mainly the author. These licenses were also used in situations of collective projects calling for a large number of participants. The system is a continuation of Copyleft[11] used for free software. In this regard, we note that the computer scientists behind the "free" software were already in a 2.0 perspective, by working remotely in small teams, sharing knowledge, and constantly challenging themselves to participate in a joint project without lucrative ambition. In this sense, software under Copyleft license may be amended; everyone can improve it and make it evolve. The success of this legal regime ought to be extended to artistic activity; however, the "free art" license was less used. According to John Samuel Beuscart [BEU 07] "the copyrighted works under this license, on the one hand, are relatively few, and ...they arise especially from very experimental artistic categories, for which there is *a priori* no commercial destiny". Perhaps through a more sustained enlightenment effort, Creative Commons licenses could quickly become popular. These will allow authors to share their digital works irrespective of the media chosen by reserving for themselves certain rights. The advantage of the system relies primarily on its simplicity: several types of "licenses", graphically notified in the form of logo on the side of the work, determine the scope of use: modified, shared, commercial, etc. In this way, authors contribute to a common heritage belonging to the "public domain" out of their own will. The Jamendo interface offered listening (through streaming) and downloading, almost exclusively under a Creative Commons license, 19,600 albums in May 2009, and it relies on 550,000 active members. This interface illustrates perfectly the transition to Web 2.0 by community (profiles, etc.) and publishing (reviews, playlists) applications, and by its economic model: it is funded by advertising and by a flat tax on donations received by artists.

12.3. Personal information property: considering the topic in the light of 2.0

The networks have rapidly become a privileged territory for unfair business practices such as spamming, profiling, etc. for more and more personalized offers, based on prospects. These practices essentially involve the collection and processing of personal information: age, profession, tastes, etc. and sometimes, political preferences, religious beliefs, union affiliation,

---

11 Copyleft is the brain child of Don Hopkins and has been developed by Richard Stallman in the GNU project including the creation of the Free Software Foundation in 1985 and the GPL was published in 1989. See "Copyleft", Wikipedia, consulted on 10 May 2009, http://fr.wikipedia.org/wiki/Copyleft.

etc. If the unique capabilities of information storage were seen as a threat to freedom previously, now it's hyperconnectivity 2.0 which arouses fears. In Europe, the protection of personal data has been recollected by the Fundamental Rights Charter of the European Union, and has been organized nationally under the European Framework Directive 95/46/EC of 24 October 1995, concerning the protection of personal data. Most countries in the world have organized a more or less binding protection. The devices, however, have never apprehend the technical and social innovations.

12.3.1. *Hyperconnectivity and personal data*

Through contributive tools and interaction capacity of information using metadata, social relationships are changing: interaction and contribution create or strengthen social ties through a community phenomenon around interests. Blogs and other applications mentioned above are concerned, but "social networks" emerge as the best example of this phenomenon. Different applications are possible: professional networks such as Viadeo or Linkedlin and friendship networks such as Facebook or MySpace.

In addition, some of the applications, which are not focused on social network, also integrate the notion of community of users. This is the case with applications such as deezer.com, which initially allowed listening to free music online, now also offers the ability to create one's profile, share playlists with other users, join fan groups, etc. The Websites such as Dailymotion and YouTube also integrate the notion of community by enabling and encouraging link between users who can comment on uploaded videos, become "friends" online, etc.

In addition to the fun applications, the connectivity offered by the 2.0 appears to be a remarkable management tool for Enterprise 2.0. Lawrence Assouad[12] describes the "ideal" global offer for "manager 2.0" thus: "while I write a mail to an employee, the service offers me the latest posts from my RSS feeds which are related to what I am writing. After reading a post from my RSS feeds, the service provides me with tweets of my followers who are discussing this topic. By consulting the identity card of a customer, I also discover his lifestreaming. In my diary, I can see not only the diary of my

---

12 See the blog of Laurent Assouad: http://www.entreprise20.fr/category/analyse/page/2/, available online on 10 May 2009.

colleagues but also the public diary of my professional network. And discover the people who have in their diary the same event as mine. The public documents I have created are automatically available on platforms like slideshare. And people who consult these documents are automatically included in my group of contacts: contacts suggestion. While writing a document, the system enables me to read other documents which address the same subject, written by my collaborators or persons who are completely unknown to me. All the statuses (Twitter, Facebook, Plaxo, etc.) of all my contacts are aggregated in one place and updated in real time. But to avoid being swamped by lifestreaming in all my contacts, I have at my disposal semantic filtering tools which allow me to retrieve in a few seconds who said/did what and when, or discover information that might have been missed at this time". If in the early years of the Internet, the concern was based on cookies, Java applets might have been used without the knowledge of the user. Henceforth, the concern is on the connectivity made possible by technical applications which represents a danger much more serious to personal data and privacy of individuals. Search engines have already aggregated all the information gleaned from various applications to obtain the profile of a person.

Moreover, Web 3.0 or Web 4.0 applications emanticipate themselves from the Web and the traditional interfaces, which allows for the possibilities of surveillance and renewed violations of individual freedoms.

12.3.2. *Hyperconnectivity before the "computer and freedoms" law*

The collection and processing of personal data are strictly supervised at the national and international levels. Many countries have established a "data protection" act in their territory. Generally, an independent authority, such as the French *Commission Nationale Informatique et Libertés* (CNIL), The Office of the Information Commissioner in the UK, the Federal Trade Commission in the United States, and the Personal Data Protection Task Force in Japan, is responsible to enforce that users' rights are respected with respect to the creation of computer files. From the moment when personal identification becomes possible, it is necessary to respect certain constraints.

The concept of personal data is quite broad. It includes the IP (Internet Protocol) address of the person, his licence plate number, his e-mail address, biometric data, etc. Collecting and processing personal data therefore must be subjected to specific rules. As stated in Article 1 of the French law of 1978, "IT must be at the service of every citizen. Its development should take place within the framework of international cooperation. It shall not prejudice human identity, or human rights, or privacy, or individual or public liberties".

The disrespect for personal data protection typically exposes the perpetrator to criminal penalties and even a jail term. Individuals from whom personal information is collected must be informed of: the mandatory or voluntary nature of their responses; the consequences to them in terms of failure to reply; the individuals or entities receiving information; the existence of a right to access and rectify.

In addition, any person establishing his identity has the right to question anyone who is implementing automated processes to determine whether they contain personal information about him and possibly engaging in communication if need be. The law allows everyone to be aware of information concerning him/her and to get it corrected or deleted.

There is also a right to object to processing of one's information, its exercise remains subject to the existence of legitimate reasons and moreover, it does not hold for the numerous public sector information processing. The right of objection is without just cause and free when the data are used for commercial prospecting purposes by the person in charge of processing.

12.4. Publishing Activity 2.0: liability and information

2.0 contributory activities renew issues related to the liability of both the content provider and those who exploit these dynamic applications. The creator of a blog, the host, the operator of a social networking site, the operator of a WebTV, the operator of a search engine, and that of a social bookmarking application have every reason to ask themselves about their liability in the event of possible infringements.

Many communication practices are contentious. The contributory Web naturally generates an increase in infringement of press, such as abuse, defamation, and also economic crimes. Since 2004, there has been a three times increase in the number of homophobic crimes[13]. Most importantly, the Web has become the most aggressive campaign field of influences: between commercial and political rivals.

12.4.1. *The responsibility of the contributor*

In 2.0, the user is both a receiver and provider of content. In some situations, he plays both the roles: for example, when sharing information on peer-to-peer networks or when he contributes to wikis, WebTV, or social networks through his personal space. He then answers to informational crimes that he might commit during this occasion: defamation, racism, etc.

The person in charge of the content (Web page or Website), is the one who provides the site or the page to the host. He is liable under the traditional rules of liability, to unfair competition, to the right of respect for privacy, etc. For example, copying a page from a competitor's Website by "cut and paste" is an act of unfair competition likely to engage the liability of its author.

Similarly, a company that inserts in the "source code" of its Website, the name and the commercial code of its competitor so as to be indexed unfairly on search engines is guilty of misconduct that may incur its liability.

The site creator is also liable when a phishing and harmful site is assessed through the hypertext links. Regarding its identification, in France, the requirement of prior notification was abolished by Article 2 of the law of 1 August 2000, for a new identification system for content authors whose activity is to provide access to online communication services other than "private correspondence". The system in place provides for the identification of content publishers by creating a "masthead" mail. The liability is then based on that of newspapers or broadcast media. Professionals must indicate: name, surname, and address on their Website, business name for legal entities. Non-professionals can remain anonymous but must disclose: the identity and address of their hosting provider, who may be required to submit complete and relevant information about the identity of the owners of

---

13 See Anne Guillard, "Les propos homophobes sur le Net ont été multipliés par trois", *Le Monde*, 14 May 2009, available at http://www.lemonde.fr/societe/article/2009/05/14/l-homo phobie-sur-le-net-a-ete-multipliee-par-trois_1193070_3224.html.

a hosted site. They cannot remain contented by merely providing eccentric information, so that they can escape their obligation[14].

Even now, the law "on daily safety" of 2001 requires the ISPs to retain connection data for a maximum period of 1 year for judicial inquiry purposes wherein they could face judicial penalties, in France. Similarly, they must ensure the veracity of the data enabling the identification of the user.

12.4.2. *Application of host status to 2.0 applications*

The EU Directive on electronic commerce of 8 June 2000 had established the main principles which govern the liability of the Internet stakeholders. These principles distinguish between two activities: "content publishing" and "the provision of technical means".

In principle, the access provider and the operator are exempted from any publishing liability: they do not have the knowledge of information circulating on their networks under the secrecy of correspondence. The content provider is totally liable, if he is identified.

The question becomes more problematic for professional and non-professional operators of applications for content dissemination. They should refer to the status of the host[15]. The latter applies to all applications which allow users to stream content. WebTV, blogs, social networks, and wikis are likely to see the publishing liability of their operators engaged on the basis of host liability[16]. In most Western states, the host is not criminally or civilly

---

14 See TGI Paris, 16 February 2005, *Droit de l'immatériel*, 2005-4, no. 126; *Communication, Commerce électronique*, 2005, p. 119, observation Patrick Grynbaum. Similarly, when the host is domiciled in the US: See TGI Paris, 20 April 2005, *Droit de l'immatériel*, 2005-5, actualité, no. 147.

15 For example, this is defined by law "of confidence in the digital economy" of 2004 (Lcen) as "individuals or entities that provide, even for free, the storage of signals, writings, images, sounds, or messages of any kind provided by recipients of these services".

16 Except in the case of a prior restraint, in this case, the operator may be prosecuted as editor i.e. as the principal author. See TGI Lyon, June 21, 2005, quoted by Vincent Fauchoux and Pierre Deprez, The Law of the Internet, p. 242-243, Litec, Paris, 2008. It must be noted that the host status is contingent. In fact in the case against Jean Yves Lambert by Myspace, the presentation frameworks put in place by MySpace executives allowed the judges to retain classification as content editor; See TGI Paris, 22 June 2007, Jean Yves Lafesse and a. c / Company Myspace, communication, electronic commerce, p. 143, 2007. However, contradictory decisions challenge this law, see Paris Commercial Court, February 20, 2008,

liable for the content of its services, unless it has actual knowledge of illegal information and did not act promptly to prevent access to that content[17]. In this case, the host replaces the judge by enjoying the legality or otherwise of content.

Other basis for engaging the liability of the host: the obligation to provide reliable contact information in a situation where the liability of a site hosted by it is challenged.

On the basis of this liability regime, the issue of liability of the blogger[18] is brought in for questioning. However, the blogger incurs the liability of the director of publication i.e. he may be prosecuted just like the principal author, if he chooses to report subjects that constitute a press offence. In contrast, in electronic commerce, the host status is sometimes abandoned in favor of the broker[19].

12.4.3. *Semantic Web: the application status of search engine*

The issue of liability of a search engine is sometimes sought[20] particularly, in unfair competition or parasitism. It is the legal standing of search engines that lays the groundwork for its status. A ruling by the Court of Appeal in Paris on May 15, 2002, stated thus: given that a litigious hyperlink produced by a search engine is nothing but a simple reference,

---

Flach Film Company and a. v. Google, Law of the immaterial, 2008/36, no. 1197 and TGI Paris, 15 April 2008, Omar S., Fred T., and a. c / SA Dailymotion quoted Fauchoux V. Deprez and P. (2008).

17 LCEN in France has returned to the principle of immunity of the host, which should no longer wait for the judge's request to remove malicious data, but must remove them as soon as they are noticed.

18 See Rambaud S., "Le Blog, objet de multiples responsabilités", *Légipresse*, no. 225, p. 103, October 2005.

19 See tribunal de commerce de Paris, 30 June 2008, Parfums Christian Dior et a. c/ Ebay Inc, available online on 10 May 2009: http://www.legalis.net/jurisprudence-decision.php3?id_article=2351.

20 Sédallian V., "A propos de la responsabilité des outils de recherche", *Juriscom.net*, 19 February 2000; Vivant M., "La responsabilité des intermédiaires de l'Internet", *La semaine juridique*, édition générale, 1999, I, 180; Bourgeois C., Livory A., "Eléments de réflexion sur la responsabilité du fournisseur d'hébergement", *Droit de l'Informatique & des télécoms*, 1999/3, p. 116; Olivier F., Barbry O., "La responsabilité des prestataires d'hébergement", note sous l'arrêt de la Cour d'appel de Paris du 10 février 1999, *La semaine juridique*, édition générale, 1999, II, 10,101.

such hyperlink need not be removed, thus exempting the designer of the tool. Search engine is not considered as a host. However, this solution is no longer valid in the case where the disputed site and search engine are contractually bound (sponsored link). In this case, the search engine will be held liable as the advertising company. This was the decision of the *Tribunal de Grande Instance de Paris* on February 4, 2005[21]: a search engine is therefore considered "as the owner of an advertising medium that offers advertisers to include their adverts on a pay site controlled by the search engine". For RSS feeds aggregation, the publishing of stream submits its operator to a debated legal regime between the host and the content editor. The judges in effect raise the point that there is possibility of making publishing choice but that the definitive information is not really chosen. Thus, whatever the application in question, it is the criterion of publishing choice that is determinant.

The question of liability of Internet actors refers to the delicate distinction between the information generated through technical devices such as mathematical algorithms and those from a genuine publishing choice. The border is often difficult to determine, which explains the jurisprudence.

12.5. Conclusion

In the 2.0, information status has evolved, and the paternity right is fading into evanescence works that are constantly changing through anonymous contributors guided by a selfless project. We are witnessing in parallel the owning of personal data, a real bargaining chip in the provision of free applications and works. Another aspect of Information 2.0 is the influence strategies, which may entail the responsibility of their authors, and sometimes dangerously questioning the liability of operators of applications of "collective intelligence". Therefore, we can only observe the emergence of new publishing models challenging traditional models governing the status of information and that of the author, for which legislators remain helpless.

---

21 TGI Paris, 4 February 2005, *Communication, Commerce électronique*, 2005, commentaire p. 117.

12.6. Bibliography

[ALL 07] ALLARD L., BLONDEAU O. (eds), "Dossier Web 2.0? Culture numérique, cultures expressives", *Médiamorphoses*, no. 21, September 2007.

[AND 06] ANDERSEN B. (ed.), *Intellectual Property Rights: Innovation, Governance and the Institutional Environment*, Edward Elgar Pub, London, 2006.

[BEN 04] BENTLY L., SHERMAN B., *Intellectual Property Law*, Oxford University Press, Oxford, 2004.

[BER 02] BERTRAND A., *La musique et le Droit de Bach à Internet*, Litec, Paris, 2002.

[BEU 07] BEUSCART J.S., La construction du marché de la musique en ligne. PhD dissertation, école normale supérieure de Cachan, 2007.

[BIL 05] BILLIAU M., "Contrefaçon, propriété et responsabilité", Communication Commerce Électronique, p. 11-12, September 2005.

[CAV 00] CAVES R.E., *Creative Industries: Contracts Between Art and Commerce*, Harvard University Press, Cambridge, London, 2000.

[FAU 08] FAUCHOUX V., DEPREZ P., *Le droit de l'Internet*, Litec, Paris, 2008.

[LLO 05] LLOYD I.J., *Information Technology Law*, Oxford University Press, Oxford, 2005.

[LUC 09] LUCIEN A., "La rémunération des auteurs dans l'économie du web 2.0, la réponse de l'état", *Terminal*, numéro spécial, La propriété intellectuelle à l'heure du numérique, p. 43-60, January 2009.

[MAI 08] MAIGRET E., "Accès, communautés et produits de contenu", in GREFFE X. and SONNAC N. (eds), *Culture web*, p. 127-136, Dalloz, Paris, 2008.

[MAN 08] MANOIR DE JUAYE (DU) T., *Le droit de l'intelligence économique*, Litec, Paris, 2008.

[ORE 05] O'REILLY T., What is Web 2.0? 2005, available at http://www.oreilly.com/.

[PEC 05] PECH L., COYNE M., "Une victoire à la Pyrrhus pour l'industrie du divertissement? La distribution de logiciel de Peer to peer à l'épreuve de la Cour suprême américaine", *Revue Lamy Droit l'Immatériel*, p. 6-7, September 2005.

[QUO 08] QUONIAM L., BOUTET C., "Web 2.0, la révolution connectique", *Document numérique*, vol. 11, no. 1-2, p. 133-143, 2008.

[ROS 08] ROSNAY (DE) J., *2020: les scénarios du futur, Comprendre le monde qui vient*, Fayard, Paris, 2008.

[ROS 08] ROSNAY (DE) J., "La "longue traîne" de l'énergie: l'espoir des petits producteurs indépendants", *Agoravox*, 4 April 2008, available at http://www. agoravox.fr/actualites/environnement/article/la-longue-traine-de-l-energie-l-3828 6?38286#commentaire1673838.

[SIR 05] SIRINELLI P., "Le peer to peer devant la Cour suprême US", *Tribune*, p. 796, Dalloz, Paris, 2005.

[THO 05] THOUMYRE L., "Peer to peer: l'exception pour copie privée s'applique bien au téléchargement", *Droit de l'immatériel*, no. 7, p. 13, 2005.

[VAN 08] VANDENDORPE C., "Le phénomène Wikipédia: une utopie en marche", *Le Débat*, no. 148, p. 17-30, January to February 2008.

[VAN 05] VANESTE C., Rapport parlementaire sur le projet de loi (no.1206) relatif aux droits d'auteur et aux droits voisins dans la société de l'information, Assemblée nationale, 7 June 2005.

Texts and decisions cited

Code de la propriété intellectuelle, Dalloz, Paris, 2007.

Convention de Berne pour la protection des œuvres littéraires et artistiques du 9 septembre 1886.

The US Supreme Court, Metro-Goldwin-Mayer Studios Inc. V. Grokster, Ltd. 545 U.S. 913 2005, http://www.law.cornell.edu/supct/html/04-480.ZS.html, available online on 10 May 2009.

The US Supreme Court, Sony Corp. Of America v. Universal City Studio, 464 US 417 1984, http://caselaw.lp.findlaw.com/scripts/getcase.pl?court=US&vol=464 &invol=417, available online on 10 May 2009.

Directive 2000/31/CE du Parlement européen et du Conseil du 8 juin 2000.

Engagement de Tunis, Sommet mondial de la société de l'information, Tunis, November 15 2005.

Loi "droits d'auteurs et droits voisins dans la société de l'information (Dadvsi)", 1 August 2006.

Loi "pour la confiance en l'économie numérique" (Lcen), 21 June 2004.

Loi "relative à l'informatique, aux fichiers et aux libertés" (informatique et libertés), 6 January 1978.

Traité de l'Office mondiale de la propriété intellectuelle, "Interprétations et exécution de phonogrammes" of 20 December 1992.

Travaux parlementaires de la loi "relative aux droits d'auteur et aux droits voisins dans la société de l'information" of 1 August 2006.

PART THREE

Territory

Chapter 13

Territory and Organizational Reputation 2.0

13.1. Introduction

The changing patterns of production and dissemination of information on the Web has marked the transition from a first-generation Internet to, what is now appropriate to be called, Web 2.0, a term that describes both technological developments often resulting from ascending innovations [CAR 06], new uses (through the democratization of tools and technologies which are simple to implement, often free in order to promote participation) and a Wide scope of social phenomenon. The new communication territories, founded by the digital culture and the expressive culture related to the advent of participatory Web, seem to be consecrated to new places of social practices, mobilization, and activism. Under the influence of changing patterns of technological, economic, and cultural production and dissemination of information, a new communication paradigm horizontal and producer of collective intelligence has been established.

In fact, the digital age allows anyone to produce or co-produce and disseminate information content which often targets a wide audience. The principle is simple: the receiver of information can become a transmitter, and hence, become a medium. The principle that governs the indistinctness of mediated communication in the social space is thus confirmed or even strengthened: "the recipients or the enunciators of speeches, messages, and

Chapter written by Serge CHAUDY and Lucia GRANGET.

other various forms of communication which organize their relationships are not based on a status or identity, but on the different contexts in which they are involved and the communication and sociability objectives they have, or which are set by the society in which they are" [LAM 97]. New paradigms of journalism such as citizen journalism, enterprise journalism, and crowdsourcing have appeared. The interest in public affairs and speaking in social spaces, expressed through blogs, wikis, forums, community sites, and social networks, continue to grow. The actors and participants are journalists by default, by taking part in the public space[1] which is more fragmented and societal, these individuals seem to be qualified intellectuals in Gramsci's sense[2], since they display a significant interest in community life [GRA 75, GRA 77, GRA 80]. This interest is then reflected in their commitment[3].

In the current crisis of representation of the postmodern (or hypermodern) society, characterized in particular by the immediacy of trade and the collapse of symbolic barriers [BOU 06], it is important to note the

---

1 Public space should be considered as a space for meetings and discussions, a symbolic space, and a historically constituted concept based on the formation of a common space. Public space, which is experiencing an unprecedented change, particularly with the advent of a participatory communication space, represents the place of communication [LAM 92], that is to say, places of meetings and discussions which enable the development of circulating opinions i.e. public opinions. The etymology of the sense leads us to the idea of a space for people, a space for everyone, a space of indistinctness (advanced by Immanuel Kant), devoted to social life and not to private life. Overall, it is jointly a communication space, a space of movement, rooted in a territory through urban spaces (streets, squares, stations, etc.), and public places (monuments, buildings, public institutions, etc.), a media space (communication mediated by the media, enrolled in a geographic area), a mediation and symbolic representations space of belonging to a social class [LAM 06].

2 For Antonio Gramsci, all men are intellectuals but not all of them occupy this social function; however, every intellectual serves a class and helps it to generate its awareness, by highlighting the cultural hegemony (maintenance means of the state in a capitalist society), the control of civil society (cultural institutions such as universities) and the process of subordination to which it can be submitted.

3 This emerging figure on the web is that of "pro-ams [ROS 06]", a new class of users of digital networks capable of producing, distributing, selling non-proprietary digital content... "Professional amateurs" (or pro-ams), they use tools similar to those of professional and that are easily accessible on the Internet. "Professionalism and amateurism, because they can be defined as 'a special relationship to an object, a way to understand and define ourself in relation to something', are the terms closely linked particularly by 'the way know-how are produced'" [RUE 93].

participatory nature of these new practices, the growing number[4] of these "web proletarian" actors [ROS 06] and their power of influence. Thus, to the principle of indistinguishability which builds a certain ideal of public space [HAB 68], one must add the power of influence on the mediated communication, which takes place across all areas of social life and by the mediation of all stakeholders of public space [LAM 97].

Organizations (considered in a broad sense: political organizations, businesses, local authorities, institutions, etc.) have understood the stakes involved in these new practices and have quickly tried to adapt and increase their visibility on these social communication spaces without always adopting an ethical approach. Indeed, Web 2.0 reflects above all a state of mind, advocating an ideal communication based on the values of sharing and exchange, and freedom of expression: the logic is called adhocratic i.e. non-hierarchical, decentralized, and often informal, unlike the structure of a traditional organization.

This logic associated with a changing discussion space has resulted in the emergence of new communication strategies in which organizations and territories are fully involved, both as objects and actors. We will first consider new strategies for organizational communication in a 2.0 environment and later address the opportunities related to the valorisation of territories.

13.2. Communication strategies of organizations in the 2.0 concept

Web 2.0, which translates into "the passage from interactivity to interaction and thus contributes to building networks that are no longer based on information sharing but the sharing of knowledge" [QUO 09], is characteristic of the development of trade-related online services; the promotion of artists, ideas, information, technology; and it relies on the viral distribution of communication that helps create a buzz. The principle of buzz, which is based on viral marketing, is that communication mobilizes all channels to create or enhance the reputation of an offer and positions the receiver (consumer) at the core of an engaged word-of-mouth process. The

---

4 Facebook has 200 million users worldwide (April 2009), where 22 million French Internet users have connected to a social network in 2008 (comScore, "Social networks are booming in France with an increase of 45%", ComScore Press Release, February 2009, http://www.comscore.com/press/release.asp?press=2725.

receiver, in turn, becomes a message sender. The 2.0 trend brings with it new markets based on new economic models, sometimes hybrid, and which seems to place "the individual on the street" [CHA 09] in the same plane as the expert[5].

13.2.1. *Web community and propagandist*

Web propagandist strategies are intended to make the best publicity of decision-making, by stimulating receptors commitment, and social mobilization by reducing the production of information to an act of propaganda, a technique which according to Jean-Marie Domenach, implements five formatting rules: simplification, magnification, orchestration, transfusion, and contagion [DOM 50]. A preliminary clarification of the concept of propaganda is key, since the term is loaded with meaning. The propaganda model [ACH 89] consists of inducing a sense of belonging, and not offering choices. It is in fact a form of political theology that justified totalitarianism. Etymologically, the term propaganda is attached to the Latin *propagare* which means "multiply by layering or cuttings and also by extending forward" [GER 91, GOU 91]. While the word only lays "emphasis on the transport of information to or through the mass of its receptors, the substantive propaganda has a dual ideological and methodological vocation: any action to spread an opinion, a religion, or a doctrine" [GOU 81]. Thus, we see the term as designating the specific part of the action regarding the dissemination of ideas to the public.

The parallel with the advertising seems good: in 1938, Serge Tchakhotine explained that "advertising forms have endless variations, which are often so unexpected and ingenious that they often inspire political propagandists". Tchakhotine still thinks that dynamic propaganda, sometimes violent, is compatible with respect for moral principles which are the foundation of the human community. Propaganda raises a paradox of democracy. Democracy provides for freedom of expression, whereas the goal of propaganda is to eliminate the possibility of choice, giving the illusion of an agreement between the propaganda and its victim. Many

---

5 The expert is the photographer, artist, journalist, political columnist, a football team coach, the bookmaker, the creator, the designer, the publicist, the producer who as a volunteer (more often) or for money can provide content, photographs, information, etc.

authors[6] agree that these instruments are neutral and that the manipulative character is not the result of techniques, but of intentional manipulation. In the celebrated phrase of the American sociologist Harold D. Lasswell, propaganda is a "mere instrument that is neither more moral nor more immoral than the handle of the water pump. It can be used for good purposes as for bad". Propaganda, as we understand it, is an ubiquitous phenomenon of our society, which uses it in advertising, politics, and therefore, distinguishing between "legitimate convincing and manipulating is not very easy": "to argue, is it not exercising a form of power over the speaker, and is it not a indirect way of influencing or manipulating?" [BRE 98].

We do not wish to discuss here issues relating to the intention of the transmitter, the lack of transparency, and the manipulative nature of the communication. We would simply define a method of presentation and publication of an opinion in a way so that it is widely publicized, both on the Web and in the traditional media, so that it can be as convincing as possible. The diversion of Jean-Marie Domenach's five rules of formatting propaganda, for use on the Web is so relevant and will rely on the construction of a hypertext and cognitive path for the user [BOU 05, HEI 08], where we would try to convince the user to adhere to the ideas issued and which will constitute the best relay of this information. Through several 2.0 devices, a rich media strategy (or rich media: videos, images, text, etc.) and several methods of influence, organizations will orchestrate the spread and influences of the messages.

13.2.2. *Web 2.0 and influence*

As part of an argumentation strategy, organizations will work on both the content of the arguments (opinions) and the container i.e. the argumentative mold that will give shape to the proposed information [BRE 96]. Initially, organizations will, from a family of arguments from the manner in which the realities are framed (case definition, authority, evidence, values and rhetorical places) and even from cropping up in reality (definition, presentation, association, dissociation), make a change in the context of reception and establish a relationship between context of reception and the opinions proposed through (deductive, analogical, etc.) arguments. If the

---

6 For Noam Chomsky, propaganda is for "fabricating consensus" for "American imperialism" [CHO 03].

organization does not seek to argue or share information but to impose it, it can also rely on a strategy of manipulation by constructing an image of reality which is not a reality (the message, in its cognitive dimension or affective form, is then designed to deceive the receiver [BRE 98]); or on a strategy of disinformation. Disinformation is an action of making a receiver, which you want to deliberately mislead, to validate some description of reality in favor of the issuer by presenting it as secure and verified information. The false information is then transformed by the construction of signs of truth into truthful and credible information.

Indeed, the social Web, by its structure – the viral nature of communication (word-of-mouth, buzz, etc.) and by the fact that technologies used in a personal context are increasingly used in a professional context, has a strong power of valorization or depreciation; this, even if only a minority of Internet users has the ability to directly influence the mass of Internet consumers [GRE 08][7]: "influencers" or "addicted creative". Thus, in implementing a 2.0 communication strategy, organizations must take into account the specific logic of the participatory and community Web and the centrality of the user. This trend of social Web should be considered when designing a new product or service: launching a product very quickly, for example, before it is finalized, so as to gather usage impressions of users, fix bugs, make improvements etc. Overall, it entails co-constructing, through continual testing phases, the project directly with the end user.

In this context, construction and animation of online communities proves useful for organizations, in particular to ensure the dissemination of information for strategic purposes; to monitor the views conveyed by the Internet users about the organization, its products or services, its employees; to anticipate and manage communicative risks. In addition, to control one's online reputation on the Internet, one should implement monitoring strategies which are adapted to new devices (social networks, Websites, blogs, forums, community media). The first principle is that of listening to the conversations (Tweets) on the social media. It is vital that the organization knows the image it conveys on the social media; it should identify its image vectors (consumers or users, competitors, employees, management, etc.).

---

7 This minority (estimated at 4%), is described as "Net addicts" or "creative junkies" by the director of Risc International, and they influence the Internet consumers (1.3 billion) [GRE 08].

13.2.3. *E-reputation*

E-reputation is the digital reputation of an organization, a cause, an event or a person (or personal branding), a kind of fame, identity and digital branding conveyed on all types of online media (social networks, Websites, forums, blogs, community media, etc.) and by all types of players. Being a differentiating factor in a competitive environment, its control is essential. E-reputation helps us establish the identity of the organization; the organization's image reflected should be consistent with the overall strategy of the organization and it's positioning, where the emphasis is on monitoring, analysis, and corrective actions.

The primary objective of an e-reputation strategy for an organization is to achieve maximum visibility on the Web. The initial actions were therefore designed to maximize the presence of the organization on the most significant parameter of the 2.0 trend following the target, the country, the strategic and operational objectives, etc. A preliminary identification of relevant features is necessary to propose a policy content which is appropriate and consistent with the corporate objectives of the brand (dedicated Weblog, creative hub on Viadéo, profile and fan pages on social networks like Facebook, real-time information and links promotion on Twitter, for example). Then, organizations can build and host online communities; and create new positions in business, in political offices, in institutions and governments, in the newsrooms of media companies such as community manager or Web journalist. Because of rapid changes on the Web as well as new uses and technological innovations, the organization is able to identify the major players in social networks, and as well identify and prioritize the information that would interest them and enlighten Internet users.

The resulting discourse on the organization (products, services, etc.) must be co-constructed with the users and in line with the marketing and communication objectives set by the management. It is essential that the image conveyed by the organization (especially through various actions and campaigns on social media) coincides with the real image of the company. Thus, the work of managing an online reputation is to bridge the gap between the image desired by the management (or ideal manage), the image perceived by the Internet users and the real image of the organization. Indeed, the digital traces persist and can have serious consequences. Moreover, the very structure of social networks tends to the inter-penetration of private and professional spheres. We thus observe an emancipation of social fields and globalization of social space: the participatory and community Web develops its own codes, references, logic and hierarchies. The expressivist phenomenon[8] linked to the success stories repeated on the Web, the increasing participation of users, and the emergence of a social Web of employees (who will depict their moods, thoughts or reviews) are strengthened by the advent of digital extimacy, where the concept of digital intimacy was dear to Lacan, which he apprehended as the actions of individuals who get their identity outside of themselves. Intimacy is the subject of overexposure and is part of a search for external legitimation of identity. This applies to the pro-ams who become experts or organization which becomes 2.0.

To build trust, organizations go beyond their markets and are engaged in this extimization process (via blogs, social networks, etc.) to consolidate a corporate image, embody the image of his organization and generate attachment. Thus community managers (CM) work on the reputation, image and identity of an organization. To build trust with the community, the CM will have to leverage the community through influencers (prescribers, opinion leaders called convicts web). The idea is that they take the information and largely relay it (via RT or Twitter relays, for example), thus prolonging it in time, and increasing visibility and audience of the message. The views conveyed are very important, it also important for the organization to be transparent, human, listening, selfless and authentic[9].

13.2.4. *Management of communities*

The instant Web which is based on word-of-mouth, leaves digital traces, and seems to supplant the importance of advertising, marketing, media relations, without completely replacing them. The principle adopted here is one of confidence, where friends (on social networking sites) give their

8 The expressivist phenomenon on the Web echoes the original claim of U.S. autonomous media driven by Jello Biafra, Dead Kennedys punk singer: "Do not hate the media. Become a media!" [ALL 07].

9 The most significant and non-compliance case with these standards is that of political candidates who a few months before an election, acquire Web participatory tools (the most common is the Weblog), and immediately suspend publications in the wake of the election date. Conversely, we note that new trends of (Web) marketing policy were commendably initiated by Barack Obama's team for the last U.S. presidential elections.

opinions on new products such as film, work, new services; by definition we trust these friends, so what they say is more important than any form of advertising because they are deemed authentic[10]. It is thus a form of néoviral community communication, opting for a non-profit tone conveying the satisfaction or dissatisfaction of the users (consumers, customers, constituents, voters, etc.). Breaking with traditional marketing techniques, the principle is to bring to the community, value, knowledge, ideas sharing and in short, resources constructed through interaction[11].

In this design, it is useful to find and create a space of expertise and skills on the Web, in other words one (or more) niche(s) that will serve as the basis for building a virtual community. We then find, in the construction of a digital social network, a growing interpenetration of private (friends, family, opinions, etc.) and professional spheres which become public. By providing value to users through information, sharing and exchange, we will gradually unite more or less active members of the future community around this project: the principle is simple; it is to bring the value-added information for benefit in return [12].

Mastering digital reputation of an organization at this social Web age and in real time depends on implementing a crisis communication strategy with the tools and technologies of Web 2.0. Whether it's a virtuous circle of communication or conversely a vicious circle, the community will play its role and through a word-of-mouth process will spread positive or negative

---

10 Seth Godin, blogger and author of influential American Tribes: We Need You to Lead Us, estimated at 1,000, the minimum number of friends or fans needed to take off a company or an artist, since 1,000 friends who speak to 1,000 other friends who in turn talk to a further 1,000 friends and so on is more effective than advertising.

11 This is to respond to all comments, even if they are negative, since it shows that we give attention to users and to the community, which also helps to boost it. It is essential to listen to community members in real time and this, constantly (real time Web), to sort the feedback, analyze it and identify true friends (or real fans) which are the best image vectors of the organization or project. Indeed, the consumer has evolved with the now immediate society. A new type of user has emerged on the Web who a company finds very difficult to manage: steeped in a culture of "free", he wants to get everything for free, and protests against advertising that is deemed too intrusive.

12 It is crucial not to solicit or attempt to sell initially, only the sharing counts (the sharing of ideas, articles, videos, experts about, links, etc.). One then has to be reactive and occupy the communication space by seeking to assist the user in his search for information. Since communication is personalized, direct, humor appears to be a big boon. Therefore, the humorous videos are plentiful on the Internet and are distributed virally.

views, favorable or unfavorable comments, on the organization. Anticipation is also a key success factor of a community communication campaign, since it can stem the phenomenon of viral communication at the slightest indication or gesture, before it spreads out. In fact, the management of communities entails monitoring activities, maintaining dialog with one's audience, listening to them and providing them with tokens of gratitude and marks of respect, and using their feedback to improve the offer. "Creating links, more than goods" is a fundamental principle of tribal marketing which finds its meaning on social networks under a different qualitative and with much lower costs.

In fact, it is clear that there are many similarities between tribal marketing and communities management: the communication tone is resolutely non-commercial, where the actors observe, interact, and share knowledge, and this helps to understand users' expectations (customers, consumers, citizens, etc.) better, create links between organizations (brands) and their audiences; leverage the visibility of the organization, etc. This digital neo-tribal marketing, which aims to inform, love, and act through community media strategies, is now used in many areas and for many purposes (media, politics, technology, personal branding, etc.). We will discuss how this type of marketing can be used for promoting territories.

13.3. Promotion of the territories

Territories, which are organized on both a physical space (geographic) and a virtual (digital) space are also enrolled, as in commercial or political organizations, in competitive logics, and in fact implement strategies of differentiation and promotion of the territory. This may be done by specific e-government policies[13], e-tourism policies, favorable tax policies, the linking of innovation actors (variables taken into account in the location strategies of firms), or the creation of participatory and community devices. Thus, the search for territory attractiveness concerns various areas; however,

---

13 For example, the SPOC (Single Online Procedures for Cross-Border Services) project aimed at improving eGovernment services to facilitate trans-border trade. Co-financed by the European Commission, and launched in cooperation from Germany, France, Greece, Italy, the Netherlands, and Poland (the project had extended to 27 states by the end of 2009) or the i2010 European Society Informations program (launched by the EU in 2005 to audit some 14,000 online services offered by public agencies in 31 countries).

we will focus our discussion on the resources built from the territories in a 2.0 environment.

13.3.1. *New areas of political communication and lobbying 2.0*

The notion of territory is considered here as a place of communication [LAM 92], where communication is conducted by multiple identities. Web 2.0 can be a tool for promoting the territories, on condition of adopting its principles and placing the user at the core of the information system. The initial experience of placing the individual at the core of the electoral process was the strategy of Barack Obama and his team during the last presidential campaign of the United States of America.

13.3.1.1. *Obamania: the first 2.0 electoral campaign*

The originality of this campaign is the perfect match of political communication with the 2.0 mindset, participatory and community, as well as convergence and coherence between the communication techniques used and the message advocating change ("I'm asking you to believe not in my ability to bring about real change in Washington ... I'm asking you to believe in yours"). Thus, most 2.0 systems were used in order to optimize the dissemination of information, to achieve maximum visibility, to establish a real exchange and reach the widest possible audience.

We note in particular the use of a campaign blog[14] launched well before the election: Rich Media. It combines writing, video, sound and image. Syndication (RSS news feed subscription to the site) can inform users in real time of the publication of a new post, for example. A battery of technological devices[15] is added to the campaign blog and allows a real policy of hyperlinks to maintain the audience and create, through the devices, a community of followers. Unlike many campaign blogs, it was the blog of transparency and was mostly fuelled by his team and his community (social network) who signed the notes systematically. The search for

---

14 A blog allows disseminating information and communicating without passing through any media filter; one can work specifically on certain points according to different strategic goals: working out the arguments of competence, expertise, authority, making a candidate more friendly or more accessible, exposing the values of the candidate (family, sports, etc.).

15 Barack Obama is everywhere: Facebook, Myspace, YouTube, Flickr, Digg, Twitter, Eventful, LinkedIn, BlackPlanet, Faithbase, Eons, Glee, MiGente, MyBatanga, AsianAve, DNC Partybuilder, etc.

closeness to the Internet users was predominant on the blog, which was an excellent entry point for the campaign team for offering rich content and new modes of communication. Added to the Weblog is a complete official site that breaks with the image of a campaign that is focused on a party: the campaign and the site are oriented toward the needs of the voter, who is positioned at the core of the device. However, the most significant digital campaign device in terms of change and 2.0 culture is the My Barack Obama site (www.mybarackobama.com) that functions as a digital social network, by providing online collaborative tools for supporters to convert them into activists' assets, raise funds[16], gain their voice, etc.

The idea was to increase communication media and work on new targets, particularly the youth, by finding an entry point to progressively encourage their active involvement on their part more, either through Facebook on which Obama had more than 3,188,000 friends, Flickr (photo album online) which, since February 2007, has 53,000 photographs of the Obama campaign published by his team or by the Internet enthusiasts[17], Obama Channel (created on YouTube in which more than 1,800 videos have been released [18]), Twitter in real time to inform the community of followers on the agenda and the logistics of the campaign. These devices have enabled reaching people who are usually disengaged and distant from politics, to gradually involve, inform them on the agenda (including through the use of SMS), to rally them to the cause, to unite them; with the ultimate goal to get their ballots.

The Obama phenomenon has been emulated in policy and strategy communication. As a good example, we may cite the copy strategies (slogans, using Internet and the 2.0 tools, mobilizing activists, etc.) implemented by the Israeli parliamentary candidates[19] or through new trends of lobbying and advocacy during the 2.0 era.

---

16 The site enabled raising a $200 million grant for a cost of $1.1 million.

17 The number of publishable photographs on Flickr is unlimited. Flicker has significant traffic and allows retrieving photographs and publishing them on another site as well as adding legends, explanations, notes issued by the public, etc. to the photographs.

18 The channel has 143,000 subscribers, the most popular video was viewed five million times and has generated 10,000 comments.

19 See in particular an article of 6 February, 2009 by Tatiana El Khoury, http://www.france24.com/fr/20090206-campagne-legislatives-israeliennes-Internet-strategie-obama-shass-likoud-kadima-plagiat.

13.3.1.2. *Lobbying 2.0*

The questioning of the new territories of political communication, brought about by social media, highlights new ways of lobbying that now seem to be exercised by the people. Lobbying is thus no longer exclusively the business of the elite but everyone's business: enlightened citizens who, from the 2.0 technologies and devices, can make their voices heard and obtain more transparency from their elected officials. Indeed, the use of ICT and 2.0 for participation and democracy tends to be normalized and democratized.

In this sense, the Congress Tweet[20] site stems from a grassroots effort, a sense of crowdsourcing to get members of Congress who would open up and communicate directly with citizens. This site lets you search for the U.S. congressmen who are present on Twitter and find them from the location data of the user (U.S. Zip Code), their addresses, their names, etc. The site presents a Tweetstream i.e. a list of recent tweets of registered members of Congress; it proposes a list of statistics showing the Senators with the strongest number of followers for the last seven days, or thirty days, besides a list of the most-watched (those with the largest community of followers)[21], a ranking of the most active, etc. A "Community" tab provides the latest tweets of the community, of "intelligent people" in dealing with the policy of the Congress or Senators in their micro-blogging activities, relaying the idea that the people are the government. A "Parties" tab shows the total number of Senators by party using Twitter (174 in total including 57 Democrats, 101 Republicans and no independent) or those not using Twitter (260 Democrats, three independents, 118 Republicans). A "Map" tab provides a map of the United States with the number of politicians using Twitter (state by state e.g. California 14/55).

This places some pressure on those who are not present on the micro-blogging site, even if only for a matter of image, particularly with the ranking of states best represented[22] (Maine 3/4).

---

20 "We the Tweeple of the United States, in order to form a more perfect government, establish communication, and promote transparency do hereby tweet the Congress of the United States of America"; http://tweetcongress.org/.

21 John McCain leads with 1,145,692 followers. A follower is a user who has signed up and registered to follow your tweets.

22 Maine tops the list with three out of four senators on the site.

The site features photographs, videos, a blog that provides information, and advice (such as the proper use of hashtags for example that enable tweets to be retrieved, and appear in the Community site). This is an opportunity for citizens to communicate and interact with their elected representatives, and an opportunity for the Senators to communicate with their constituencies. This device is therefore designed to make democracy and the exercise of power more transparent. In fact, many jurisdictions have been inspired by this device by establishing a similar connecting sites among which we find TweetCommons which links Canadian citizens to their elected; Twitica is the Portuguese version of TweetCongress; TweetMP is the Australian version; the Swedish version is Twixdagen, while the Danish is Twittertinget; EuropaTweets is the European version (What is your parliament doing?).

New forms of lobbying on this basis have emerged as the tool of civic engagement "Tweet Your Senator", accessible from the site "Organizing for America". The objective is to stimulate political accountability of Senators from twittering lobbyists who are able to speak directly to the Senator of their constituency.

The Sunlight Foundation[23] (advocating the use of the revolutionary power of the Internet to make information on the Congress and the federal government more accessible and meaningful to citizens) whose slogan means "Join the transparency movement" has launched a crowdsourcing lobbying initiative by inviting Twitter users to massively ask the present Senators on Twitter to adopt the S482 law which, for the sake of more transparency, would require Senators to electronically disseminate their campaign accounts[24].

A global phenomenon of collective and competitive intelligence is under way. More than a documentary film about the power of crowds, a film project about the power of mass collaboration, government and the Internet by Ivo Gormeley titled "Us now" produced by Banyak Films, discusses democracy, by adopting the principle that the Internet users can collectively

---

23 http://www.sunlightfoundation.com/.

24 Note that the U.S. government has committed itself for the past several months to Public Diplomacy 2.0. In lobbying for the U.S. healthcare system reform, Obama had mobilized influential bloggers for their support for the reform and enabled them to massively relay information, thus bypassing the traditional media which is rather hostile to the reform concept.

manage a complex organization (such as a football club) and explain how citizen participation through social media, could transform the way power is exercised and the way by which countries are governed.

In fact, new technologies and increasing acculturation to online collaboration enable the emergence of new patterns of social organization. For its projection in Britain at the House of Parliament (July 6, 2009), users could retrieve a sample letter on the blog project to invite their representatives or Members of Parliament (MP) to attend the screening of the film and discuss the future government. To communicate about the film, the project relies on civic engagement and the power of the masses advocated in the documentary[25].

13.3.2. *Digital Territories 2.0: state of the art*

Generally, Internet portals of regional communities are not a social place for the exchange of information and services. They must adapt to provide a dedicated area (apprehended in a 2.0 political dimension) of customizable services, easy access, generating social ties, while animating the local life. Above all, they must become tools of local and participatory democracy by establishing a real dialog with the citizen. It is essential that the platforms designed with a 2.0 target should not be too focused on technology but must be user-centric; if the contrary is the case, the information will remain ascending and the re-appropriation of the device by users will almost be zero.

While in the 2.0 culture, technologies are supported by, and evolve according to, their uses and social functions, regional communities remain in a 1.0 step where the quest for participation materializes on corporate sites through a free expression space, which is little or not visited. Digital territories, like physical territories, must always be re-appropriated by their inhabitants (users). This happens on the Internet through technologies that enable the proliferation of the capacity to produce, participate, co-construct and disseminate information by citizens, who, through the establishment, by the territory, of a single platform of valorization and aggregation of existing tools, services, experiments, will stimulate social innovation.

---

25 The film was also screened at the Hague, in Germany, in France (projection at the National Assembly on 22 June 2009 and organized by the 27th Region, public agency for innovation) etc.

The link between physical and virtual territory must be strengthened, in other words, information on the Internet must be anchored to the physical territory. Indeed, the finding of linking the information to regional contingencies is easy: both the production and the reception of information influencing the news journalism, often respond to the famous "law of proximity" (here and now), analysis centered on the ego drives of readers. The device (blog, Website, etc.) which can be accessed from around the world finds that its audience is generally limited to a restricted community (in a neighborhood, a group of friends, etc. (also language, practice)). "Curiously, in fact, one of the most amazing and paradoxical effects of a network intended to facilitate the explosion of global exchange, is to generate a significant return of territory (with limited vision), particularly at the local and regional levels, sometimes that of an ethnic, linguistic, or religious community" [PEL 00].

Social innovation must lie at both the technological level and the level of services and participation. The services must be associated with a social function, a social utility, facilitating the activities (such as trade, for example) on the territory and be consistent with social issues (environment, employment, education, economic development, culture, etc.). Nomadism is a Web 2.0 feature, and is therefore a territorial digital platform of services, where Information 2.0 should enable its users to be liberated by their computers, via applications for third-generation mobile phones, for example.

13.3.3. *The 2.0 stakes for the territories*

The stakes and risks of 2.0 communication are similar for the organizations and territories. Sharing and dissemination of information, increasing their visibility, optimization of image capital and control of e-reputation, adaptation to the context of freedom of expression of the user, seduce both organizations and local communities (concepts like cities 2.0, emerging regions 2.0, etc.). Indeed, people communicate differently since the democratization and development of digital social media have impacted the society as a whole.

The elaboration of dedicated strategies and of information, communication and technology process (protocols and procedures) are under development and aimed toward the spread of a common culture, shared values and renewal of the regional ethical framework. In addition, they encourage decision-making in social networking environments. These community devices of territorial intelligence aim at "orchestrating the cooperative analysis of data, interpretation of results and finally citizen participation in decision-making" [GIR 07]. Control and preservation of strategic information is key for competitiveness and development. In the context of the social Web, it therefore seems appropriate, in order to assist in decision-making, to provide the territorial organizations with tools to understand how to identify, select and analyze information, distribute it and then receive feedback from the receivers and users.

Territories must be able to stimulate territorial dynamics by fostering cooperation among heterogeneous actors, pooling and thereby enhancing the built territorial resources.

Online social networks and community spaces allow creating and maintaining social ties virtually, and hence contribute to the acceleration of trade. Thus, they are gradually being considered by the territorial organizations in implementing and strengthening strategies that contribute to establishing a social relationship of proximity.

In the implementation of such a participatory information and communication approach, territories have the ambition of creating innovative content that may encourage membership and participation of heterogeneous actors around newly initiated projects. As in the construction of communities, the system relies on devices creating new social spaces that enable knowledge sharing and dissemination of strategic information.

13.3.4. *Two laudable initiatives: the first regional digital social network in France and social network of the Ministry of Culture of Brazil*

The first French region to be really adapted to this environment is the Auvergne region (first French region in terms of broadband), which launched in May 2009 its social network, real space for exchange, sharing and freedom of expression, destined to all true Auvergnians[26].

In addition to the general features found on Facebook and Twitter, the site hosts a contributory WebTV and editorializes the content to highlight relevant information. Officials of the territorial social networking site

---

26 www.auwwwergne.com.

aggregate content related to regional news and get them from member blogs. The tone, specular proxemics vector, is decidedly offbeat; a special place is given to humor with "wie-dung" for example, the Auvergne equivalent of the large audience site Vdm (www.viedemerde.fr, sharing site of small mishaps and funny stories of daily life). In this sense, territorial information portals (promotional tools) will increasingly take the form of social networks of proximity[27], organizing social relationships and enriched with many services to individuals, companies, associations, traders, etc. The establishment of such an approach for territorial promotion must meet several objectives: facilitate the participation of every one in local life; create social bonds among all local actors, communities, generations by promoting conditions of interaction and by linking the providers and requesters (of jobs, organizations, services, etc.); invigorate the physical territory, economic and social development; promote mobility support; develop solidarity, uniting people around a common identity.

Brazil has rapidly deployed a device to develop a new approach to public policy and culture with the portal http://www.culturadigital.br/. Indeed, the Brazilian Ministry of Culture launched on 31 July, 2009 an open and participatory Web platform in the form of a digital social network that gives pride of place to a digital forum that revolves mainly around five themes for discussion (though not restrictive): digital art and new technologies, digital communication, digital economy, infrastructure dedicated to digital culture and digital memory.

The territorial social network of Brazilian digital culture aims to change the public space by promoting openness, the democratic formulation and construction of a public policy of digital culture by improving the conditions of interaction among its different public (government institutions, government, civil society, market culture, citizens, etc.) and by stimulating the exchange as well as co-production of events and cultural and citizens projects.

We note that in addition to traditional social networking features (creating profiles, publications, information, links, photos and videos, etc.)

---

27 A recent tender of the State on innovative web services is to stimulate and support initiatives for local government 2.0, http://www.telecom.gouv.fr/rubriques-menu/soutiens-financements/programmes-nationaux/volet-numerique-du-plan-relance/services-innovants-du-web/396.html.

a blog is added, which the users can receive through registration. The Brazilian site supports a wide range of cultural programs on the basis of citizen initiatives. At the core of the concerns is another reflection of how technology can contribute to the improvement of social life.

13.4. Conclusion

In the era of community and participatory Web in real time, it seems crucial that organizations and territories manage to adapt to new methods of production and dissemination of information and to the viral communication context of the social Web, characterized particularly by the interpenetration of private and professional spheres and information risk. Understanding, monitoring, anticipating and acting are the key words to ensure control of reputation, identity and image of the organization and the territory. Territories and organizations should keep in mind that technology adoption has often been driven by personal use[28]. For successful transition to Territorial Organization 2.0, it is crucial not to break the habits of users and to respect the values promoted by the participatory and community Web (the collaborative, knowledge sharing, etc.). To this end, it should be recollected that a regional social network has to offer three main social objects to users: socialization, networking, and social navigation [THE 09].

In fact, the experimental phases conducted on these topics show that it is imperative that the devices and technologies are subject to re-appropriation by their users. For this, they need to provide more in terms of use, they should facilitate communication and exchanges between members of a community (among citizens, employees, brands and their audiences, elected officials and their constituents, etc.) by pledging more transparency. Services must be customized and personalized. For a territory, it entails providing a digital social space of personalized answers to everyday needs, an extimization space that shapes and strengthens identities by building relationships with each other. The device must contribute to the creation of social ties, fostering close relations, strengthening regional identities, providing social and collaborative tools and stimulating adhesion to projects. The management of communities, both for an organization and a territory, should continue to gain importance. In this sense, numerous bids of local

---

28 Social innovation often comes from users. In the sense, user-driven projects applied to the environment, education, social and citizenship are numerous.

community, state, Europe, and companies indicate a growing interest in the social Web.

Increasingly, communication in organizations is considered in a cross-media dimension (or media crossing). On the Web, this means multiplying channels of information dissemination devices and combining all the resources of the social Web to increase its visibility: corporate Websites, blogs, social networking sites and fan pages, sites content sharing (such as Slideshare, Digg or Del.icio.us), media contributions, community media and citizen journalism sites, the contributory encyclopedia: Wikipedia, etc. This trend of convergence of communication around a common goal applies to at all levels of social life (economics, information, politics, etc.). The outcome of this is a usage oriented toward promoting events, territories, businesses, causes, political candidates, media (strongly present particularly on Twitter).

13.5. Bibliography

[ACH 89] ACHACHE G., *Le marketing politique*, no. 4, Revue Hermès, Paris, 1989.

[ALL 07a] ALLARD L., "Emergence des cultures expressives, d'internet au mobile", p. 19-25, in L. ALLARD, O. BLONDEAU (eds), "Dossier Web 2.0? Culture numérique, cultures expressives", *Médiamorphoses*, no. 21, September 2007.

[ALL 07b] ALLARD L., BLONDEAU O. (eds), "Dossier Web 2.0? Culture numérique, cultures expressives", *Médiamorphoses*, no. 21, September 2007.

[BOU 05] BOUTIN E., ROMMA N., "Les stratégies d'influence sur internet, validation expérimentale sur le lobby antinucléaire", *V° colloque de la SFBA* (Société Française de la Bibliométrie Appliquée), Ile Rousse, June 2005.

[BOU 06] BOUGNOUX D., *La crise de la représentation*, La Découverte, Paris, 2006.

[BRE 96] BRETON P., *L'argumentation dans la communication*, La Découverte, Paris, 1996.

[BRE 98] BRETON P., *La parole manipulée*, La Découverte, Paris, 1998.

[CAR 06] CARDON D., DELAUNAY-TETEREL H., "La production de soi comme technique relationnelle", *Réseaux*, no. 138, p. 15-72, 2006.

[CHA 09] CHAUDY S., PELISSIER N., "Le journalisme participatif et citoyen sur internet: un populisme dans l'air du temps", *Quaderni*, no. 70, Editions de la maison des Sciences de l'homme, Paris, October 2009.

[CHO 03] CHOMSKY N., HERMAN E.S., *La Fabrique de l'opinion publique, la politique économique des médias américains*, Le Serpent à Plumes, Paris, 2003.

[DOM 50] DOMENACH J.M., *La propagande dans tous ses états*, PUF, Paris, 1950.

[GRA 75] GRAMSCI A., *Ecrits politiques II, 1921-1922, Textes choisis, présentés et annotés par Robert Paris,* Gallimard, Paris, 1975.

[GRA 77] GRAMSCI A., *Ecrits politiques I, 1914-1920, Textes choisis, présentés et annotés par Robert Paris,* Gallimard, Paris, 1977.

[GRA 80] GRAMSCI A., *Ecrits politiques III, 1923-1926, Textes choisis, présentés et annotés par Robert Paris,* Gallimard, Paris, 1980.

[GER 91] GERVEREAU L., *La propagande par l'affiche, Histoire de l'affiche politique en France, 1450-1990*, Editions Syros-Alternatives, Paris, 1991.

[GIR 07] GIRARDOT J.J., Système communautaire d'information territoriale, UMR 6040 Thema: Théoriser et Modéliser pour Aménager, 2007, available at: http://thema.univ-fcomte.fr/Systemes-communautaires-d.

[GOU 81] GOUREVITCH J.P., *La propagande dans tous ses états*, Flammarion, Paris, 1981.

[GOU 91] GOUREVITCH J.P., lecture notes, "La propagande par l'affiche, Histoire de l'affiche politique en France, 1450-1990", in L. GERVEREAU (ed.), *Revue française de science politique*, vol. 41, no. 4, p. 582-583, 1991.

[GRE 08] GREEN S., Dynamiques d'influence sur le web, Risc International – White paper, 2008, http://www.risc-international.com/images/upload/portfolio_img/web%20WHITE%20PAPER.pdf.

[HAB 68] HABERMAS J., *L'espace public*, Payot, Paris, 1968.

[HEI 08] HEIDERICH D., *Influence sur internet: Perceptions et mécanismes d'influence sur internet dans la société de l'urgence,* Observatoire international des crises, 2008, www.communication-sensible.com/download/influence-sur-internet-didier-heiderich.pdf.

[LAM 92] LAMIZET B., *Les lieux de la communication*, Collection Philosophie et langage, Margana, Liège, 1992.

[LAM 97] LAMIZET B., SILEM A. (eds), *Dictionnaire des sciences de l'information et de la communication*, Ellipses, Paris, 1997.

[LAM 06] LAMIZET B., *Sémiotique de l'événement*, Hermès, Paris, 2006.

[PEL 00] PELISSIER N., Les mutations du journalisme à l'heure des nouveaux réseaux, http://www.diplomatie.gouv.fr/fr/IMG/pdf/FD001421.pdf.

[QUO 09] QUONIAM L., LUCIEN A., "Du web 2.0 à l'intelligence compétitive 2.0", *Actes du colloque ISKO*, Lyon, France, 2009, http://sites.google.com/site/master2ietustvhanoi/publications.

[ROS 06] ROSNAY (DE) J., REVELLI C., *La révolte du pronet@riat: des mass média aux média des masses*, Transversales, Fayard, Paris, 2006.

[RUE 93] RUELLAN D., *Le professionnalisme du flou*, Presses Universitaires de Grenoble, Grenoble, 1993.

[THE 09] THELWALL M., *Social Network Sites: Users and Uses*, in M. ZELKOWITZ (ed.), *Advances In Computers*, Elsevier, Amsterdam, 2009.

Chapter 14

Triple Helix and Territorial Intelligence 2.0

The aim of this chapter is to present the public policy of competitive intelligence in Brazil through the example of the implementation of a strategic intelligence model based on a university-industry-government relationship for the consolidation of a national system of Science, Technology, and Innovation (SNCT&I, in Portuguese). This public policy is based on the principle of bridging knowledge and skills transfer, and it relies on a Government Device 2.0: the innovation portal.

14.1. Evolution in the 2.0 world

The emergence of the 2.0 concept, first on the Web, was the realization of profound changes in the societal organization. However, more than a technical phenomenon, it is in fact the consecration of new modes of operation, organization, learning, and decision-making. This has led to the emergence of the 2.0 concept, which is commonplace today, with the use of concepts like Enterprise 2.0, Education 2.0, and Management 2.0. The question now arises: What is the phenomenon by which an activity emerges as 2.0? This phenomenon basically involves an architecture of participation, a social architecture, and an architecture of shared, collaborative and distributed computer applications.

Chapter written by Rosana PAULUCI.

In this context, the concept of competitive intelligence has also evolved [QUO 09]. From the early form of documentary watch, it has grown and become diversified and complex to achieve what is called Competitive Intelligence 2.0. This development is not always easy to understand since it requires good knowledge of the foundation and history of this activity. It also requires constant new learning, close observation of the implications of their developments, and understanding the major phenomena of society.

State and government institutions must respond to this new paradigm, by an approach, perhaps not entirely new but essential: the systemic approach, reflecting the breadth of decision-maker scope within the framework of their responsibilities, without departing from the globalized world events or commitment to society.

We can therefore consider that the function of strategic monitoring systems is essential for the inclusion and maintenance of countries in the 2.0 domain, since, in addition to specific data and the possibility of integration with other systems or databases, they also incorporate expert analysis, the fruits of the intelligence process or knowledge structuring for decision-making.

However, the function, structure, management, and implementation of the results of government information systems will be increasingly present in developing the public policy and for the adequacy of investment required for the social and economic development of countries, thus enabling Governance 2.0.

14.2. Knowledge, innovation, and development

Information and knowledge are key elements for the social and economic development of regions and countries. Over the years, given the transformation of social, political and economic practices, organizations, nations, and regions have been forced to question the institutional apparatus and organizational structures in place. The speed and depth of the changes are indicative of the intensification of the effort needed to develop in harmony with the progress of scientific knowledge, and develop social and economic strategies for the positive integration of individuals, organizations, regions, and nations.

Castells [CAS 98] defines the current economy as the era of information economy, since productivity and competitiveness of units and agents depend mainly on their ability to manage, process, and effectively apply information based on knowledge.

For all, information and knowledge are and will be always the key elements for economic growth and technological change. These two factors have largely determined the productive capacity of society and living standards, as well as social forms of economic organization.

Therefore, as noted by Plonski [PLO 95], there has been a paradigm shift from industrial society to the knowledge society, where knowledge and its management have emerged as major factors that count in organizational and national competitiveness, and are the center of debate for constructions related to social and economic development.

Meanwhile, the Organization for Economic Development Cooperation (OEDC) believes that the competitiveness and productivity of countries and their industries are strongly determined by technological progress. Studies indicate that during the process of innovation, industries seek to use a wide variety of information sources, including search sources to improve their products or processes. However, the OEDC and the World Bank recognize that the determinants of a national system of innovation are: the control and macroeconomic context; the education and qualification system; communication infrastructure and market dynamics.

These factors determine the national capacity to establish a research network of companies, to develop its scientific and technological system and, more importantly, to have a direct influence on the national potential for the creation, dissemination and use of knowledge in the innovation process [OCD 92]. The ability to create knowledge, macroeconomic space and favorable regulatory conditions, associated with state priority guidelines, are necessary ingredients to the momentum of the national innovation system.

In fact, the innovation process can be improved when we encourage the formation of increasingly intense networks between industries, research institutes and universities. Forming and inducing university-industry networks are therefore essential to the innovation chain of a country [AMA 04].

The significant increase in university-industry relationships is due to a combination of various factors, among which specialized literature highlights the acceleration of the pace of transition toward a knowledge-based economy; the development of economic globalization and competition between companies; budgetary restrictions and general reduction in research funding in almost all countries; and the high rising costs of research and development [GUS 97].

As proposed by the triple helix model [ETZ 98], in the knowledge society, the process of innovation is the result of the dynamic, flexible and interdependent interaction in a triangular cooperation process between the government, university and industry. It is the Government that performs the role of inducer and facilitator of university-industry interaction for the application of new knowledge and creating products and services.

As can be seen, the role of the state is fundamental to the promotion and intensification of this dynamic field and process which encompasses social and economic development, strengthening of national innovation systems, collaboration of networks of actors in the system for cooperation, and the dynamics of innovation itself.

If the presence of the state is required to assist and promote these interactions, it is equally essential that the state be integrated into the globalized world, in order to fulfil its duties more effectively.

In this sense, the digital inclusion of the state becomes dominant for the provision of services to the society as a whole. What is defined as "virtual state" [FOU 01], is the challenge of incorporating information and communication technology that determines how a government is organized according to virtual agencies and public and private networks, intergovernmental agreements which bind the actors at the federal, state and local level, without profit-making and private goal, and Web-based services that connect organizations.

The use of information and communication technologies also allows a better systematization of processes, which makes the government machine less slow and helps to improve the evaluation of government actions due to the existence of a structured system for collecting, processing and disseminating information.

14.3. The ST&I systems for Brazilian intelligence

In Brazil, the great transformation to the new paradigm and the incorporation of new world model national systems began in 2004, by the approval of the law on innovation [BRA 04], by setting guidelines for supporting the creation of companies that are technology-based and for interaction between the academic (universities) and the industrial (business) sectors, as well as the rules for marketing innovation produced in academia or in research and development (R&D) centers. In addition, the government developed a new industrial policy in 2004 which focused on strategic sectors for the repositioning of Brazil in the international trade arena. In this context, the expansion of specialized and cooperative spaces for innovation was strongly stimulated, through strategic alliances and development of projects involving national companies, scientific and technological institutions (STIs) and private law organizations without profit-making goal, to create innovative products and processes.

14.3.1. *Lattes System*

The first element that configured the beginning of the discussion was the development of the Lattes System, which contains data and information on technical and scientific production in Brazil. La Plataforma Lattes represents the CNPq[1] experience in integrating databases of CVs and science and technology institutions into a single information system, whose current importance extends not only to support operational activities of CNPq, but also to support other federal agencies and regional organizations.

Due to their scope, the information contained in the Lattes System can be used both to support the activities of management and the formulation of policies for the science and technology sector. The adoption of a national curriculum vitae model, with the wealth of information in this system, and its mandatory use for each funding request, as well as making these data available on the Internet, has provided additional transparency and reliability to the support activities of the agency. Lattes Curriculum, which highlights the past and current life of researchers, is an indispensable element for the analysis of competence and merit of applications submitted to the agency. From Lattes Curriculum, CNPq has developed a standard format for collecting information

---

1 National Council for Scientific and Technological Development (CNPq) is an agency linked to the Ministry of Science and Tecnologia (MCT), dedicated to the promotion of scientific and technological research and to the formation of human resources for research in the country. Its history is directly linked to the scientific and technological development of Brazil.

on personal background, now adopted not only by the agency, but also by most aid agencies, universities and research institutes in Brazil.

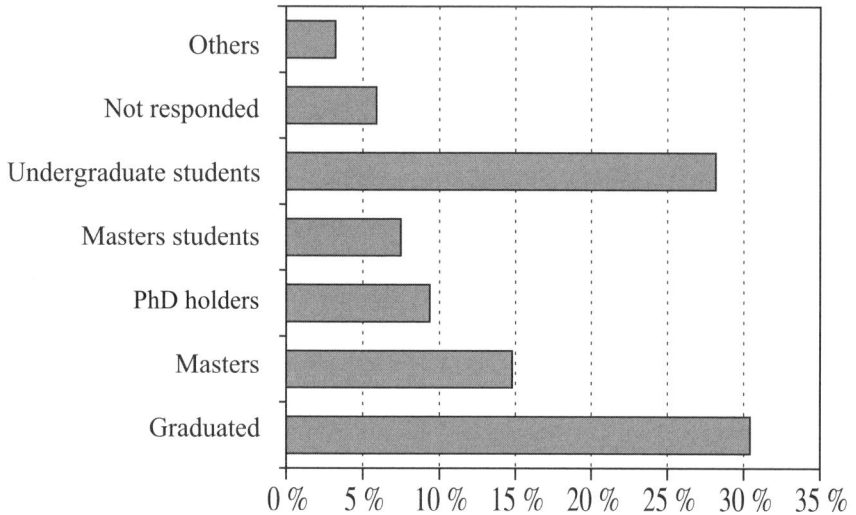

Figure 14.1. *CV database by education level*

CV based on educational level	%	Number of CVs
Bachelor degree	31%	349,227
Masters	15%	170,013
Doctorate	9%	107,356
Postgraduate students	8%	86,914
Undergraduate students	28%	321,568
Not mentioned	6%	67,961
Others	3%	39,029
		1,142,068

Table 14.1. *Number of CVs by education level*

Currently, the Lattes System database has nearly 1.1 million CVs, where 31% are made up of PhD holders, Masters and postgraduate students and 59% are made up of students and first degree holders [PLA 09].

The distribution of doctorate holders contained in the Lattes System by sector shows a greater concentration in the areas of natural and Earth sciences and humanities, followed by the health sciences, biology and engineering sectors.

The database of institutions in the Lattes System contains nearly 4,000 institutions throughout the country, spread across different sectors: education, businesses, private non-profit-making, and government.

Sector	%	Institutions
Higher education sector	22%	1,035
Technical and secondary education sector	1%	66
Company sector	30%	1,405
Government sector	14%	628
Private non-profit making sector	32%	1,483
		4,617

Table 14.2. *Institutions by sector in Lattes System*

14.3.2. *Directory of research groups*

The directory work was started in 1992 at CNPq and, every 2 years or so, the agency discloses an inventory of installed capacity for research in the country, a capacity measured by the active groups at each period, to the general public.

The directory of research groups in Brazil is an inventory of active research groups in the country. Its databases contain information on human resources constituting groups, lines of ongoing research[2], knowledge specialists, concerned sectors, the scientific, technological and artistic productions of researchers and students who are part of groups and the types of interaction with the productive sector.

---

2 The research line represents the axis of scientific studies which are based on the tradition of investigation, where the results of the emerging projects show affinities between them. The research project concerns investigation with well-defined starting and ending dates, based on specific goals, aimed at achieving results, cause and effect or the highlighting of new developments.

These groups are located in universities, isolated institutions of higher learning, scientific research institutes, technological institutes, research and development laboratories of public and private enterprises, as well as in some of the non-governmental organizations conducting research.

Censuses have presented quantitative information about groups in their various dimensions and have provided resources for text searches on databases[3]. The directory of research groups has three main purposes:

– a tool for information exchange for the scientific and technological community in the conduct of professional activities;

– a tool for planning and managing science and technology activities, since, besides the information directly available on the groups, the database is basically based on censuses. Whether at the level of institutions or scientific societies or various political-administrative bodies of the country, the database directory is an inexhaustible source of information;

– storage of science and technological activities in Brazil, since the databases contain the results of censuses and censuses are recurring, plays an important role in its preservation. The directory of research groups currently has 443 groups, representing all the regions of Brazil, especially the South Central region, which has the largest concentration of demand and supply in science and technology.

14.3.3. *Potential of systems for strategic cooperation*

Considered in isolation, Lattes System databases and research groups directory are not an aid to decision-making and strategic orientation regarding investments and the Brazilian policy on science, technology, and innovation in the public domain or private sector, particularly because the platform and the directory were not built for this purpose. To do this, it would be important to consider data from these two databases along with new cross-analyses and results.

Among key features and capabilities, we can report the identification, characterization and detail of Brazilian investments in research and development (R & D), by considering specific periods of time, large areas of

---

3 The Website (http://dgp.cnpq.br) contains the censuses from the year 2000.

knowledge, economic sectors, regions of the country, research institutes, researchers, research lines, etc. As we have observed from the databases, it would be possible to identify the scientific and technological knowledge available and put into application in universities and national research institutions, in addition to the specificity of the accomplishments.

Given the large potential of an information system that contains – or would allow – the identification of supply and demand of scientific and technological knowledge, and in a way to help increase the relationships between the engines of social and economic development of the country, the Brazilian government has mobilized its efforts. To bring the agents together, it was necessary to know the national expertise in various fields of knowledge, detect and encourage the formation of research networks with the aim of providing assistance to cooperation and innovation, while respecting, in this process, the differences, expectations and fears of each representative of each of the triple helix model entities (university, industry and government).

To this end, the Ministry of Science and Technology (MST), as part of its remit, has requested the Center for Management and Strategic Studies (CGEE)[4] to be in charge of a national project, in order to establish a mechanism to support the new demands of agents of the national system of innovation (NSI). Charged with the mission to develop a mechanism or instrument, capable of supporting and even, to some extent, intensifing the process of interaction of agents for cooperation and innovation, including public and private initiatives, the CGEE had organized a multidisciplinary team of highly experienced people in the following areas: public policy and science system, technology and innovation, university-industry relationship; innovation processes, development of e-government application. However, responses to questions relating to the strategic orientation of these partnerships, which constitute the strengthening of the NIS were not

---

4 CGEE is a social organization, associated with the MST by an administrative contract, which gives it the task of realizing a portfolio of projects considered strategic for the country, with respect to ST&I and the productive sector. Its principal functions are: (a) to promote and conduct high level studies and prospective research in the field of science and technology and their relationships with the productive sectors, (b) to promote and carry out evaluation activities of strategies and economic and social impacts of policies, scientific and technological programs and projects, (c) to disseminate information, experiences and projects within the society, (d) to promote interlocution, articulation and the interaction of science and technology and productive sectors, (e) to develop technical and logistic support activities to public and private institutions.

included with an objective of reflection and development. The initial challenge consists of creating information on the demand and supply of scientific and technological knowledge compatible by indicating the agents and needs, and is now the subject of the Innovation Portal[5].

There remains however a second challenge, that of enabling the strategic orientation of ST&I, by taking into account the data and information available in the national science and technology system, from its databases and the Innovation Portal. It should be noted that both the described databases and the Innovation Portal have the national ST&I as their reference – in as much as their main funder is the Department of Science and Technology. To address this second challenge, the suggestion is the adoption of a strategic intelligence system (still in the design phase) from the Innovation Portal: a system that will be presented below.

14.4. Innovation Portal, the observatory for strategic intelligence

The challenge for the team was to establish an initial configuration for the cooperation mechanism among agents in the innovation chain, from national and international experiences and key information sources related to the innovation theme, available in the country. From the presentation and the discussion of the first proposal, it was necessary to introduce evolutionary change in the design of the mechanism and the composition of what would be the project for the development of this system.

From the discussions it was identified that this mechanism would be an information system or platform[6], to house supply and demand of knowledge, the scope of which allows us to incorporate the interests of the industry and the university for technological cooperation.

The goals of the Innovation Portal were defined as follows: mapping skills and offers, allowing company managers and other innovation actors to know the technological skills available nationally; mapping company demands and other innovation actors which may lead to cooperation in focusing on innovation; the interaction between the supply of technical and

---

5 The Innovation Portal has been available to agents since 2005 and undergoes successive updates or enhancements requested by users.

6 The platform is understood as a set of systems, which is what actually the portal becomes, in terms of systems engineering.

scientific knowledge and the demand expressed by the business sector; support for innovation management through information and knowledge systems and their integration with other initiatives in innovation; the disclosure of instruments to support innovation among all the stakeholders in the system, in cooperation with other initiatives.

14.4.1. *Foundations of Innovation Portal*

As we have seen above, the demand for a support instrument for the dynamics of innovation, the support for university-industry-government interaction for cooperation, emerged from government action related to science and technology domain, which presupposes some of the foundations that reflects: being a government action; deciding the interaction between the university and the industry; having the purpose of innovation, which presupposes knowledge transfer. In this sense, the team of specialists involved in the design of the Innovation Portal established three conceptual dimensions, which, in a non-hierarchical way, guided the construction of the portal:

– e-government: this is a government action and that the government is an actor of the instrument. The other factor relates to the need to be associated with current federal guidelines on the integration of the country into the knowledge society and the digital age;

– knowledge management: as the effort tended toward obtaining an instrument that serves to support the innovation process, so as to intensify it, it was necessary to adopt knowledge management as a theoretical reference since knowledge is considered as the basis of the innovative process;

– Triple Helix Model: since the concern is about the university-industry-government relationship, it is essential to choose a conceptual model which represents the national innovation system and the university-industry-government relationship.

Another issue became urgent since the Innovation Portal was incorporated into the dynamics of economic development of Brazil, as advocated by the specialized literature discussed in the previous chapter. In addition, Brazil has a scrics of international commitments (specific programs and projects) related to the same economic development, but especially to the conceptual dimensions defined for the portal. The question that arises concerns the possible response of the Innovation Portal to the national goal of increasing the economic and social development, with which the government programs, projects and actions would be associated.

In response, we chose two complementary approaches to the assumptions adopted initially for the construction of the Innovation Portal, in order to take into account the dimensions of national economic development, where the developments are part of the international commitments assumed by the Brazilian government: knowledge-for-development and e-government-for-development schemes. This analysis will also indicate to what extent the Innovation Portal will additionally meet the requirements of the two new approaches in the Brazilian context.

As a support for the challenges raised, the Innovation Portal has adopted a specific architecture. The system architecture built for Innovation Portal is based on the three theoretical foundations adopted, but has its specific software engineering features. The architecture of e-government was composed of five levels which represent the technological resources used, and are defined as follows:

– standardization of units of information and primary sources, transactional information database;

– information systems;

– secondary sources: data warehouses;

– areas of presentation, including the portal;

– knowledge systems.

14.4.2. *Actors of Innovation Portal*

The initial phase of the Innovation Portal included only two groups of actors: the first relating to the productive sector (the knowledge seeker) and the second relates to the national academic community gathered in the Lattes System (the knowledge supplier), where the portal was fundamental to the technological cooperation, which would be the main objective of the Innovation Portal.

The later phases have been rewarding in that new players have realized the importance of the Innovation Portal, which is presently still in its preliminary version. These actors have requested that their groups be included in the Portal environment, since they were involved in the innovation process and are part of the Brazilian system of innovation i.e. they were already directly articulated to achieve the goals proposed to the portal actors: companies; specialists, and research and development groups; institutes of technological research and innovation and the general public.

Companies can find and provide skills and opportunities in all areas of knowledge and in different sectors of the economy. Their environment allows them access to skills, to interact with experts, to get exclusive information, make requests for cooperation and disclose their skills. Specialists, and research and development groups provide skills and find opportunities for interaction and cooperation with the business sector. Their environment allows the supply of skills, interaction with companies and access to proprietary information. Research institutes in technology and innovation monitor the process of cooperation, participate in it and find in it facilitating tools to meet the requirements of the law on innovation. Their environment allows leaders of R&D institutions to follow-up the cooperation processes, and interact with companies as well as the technical & scientific community.

Interested parties may participate in the Innovation Portal, access its content, indicators, and conduct a search for skills and opportunities for technical and scientific cooperation. The public (anonymous users) have access to the Portal's regionalized usage indicators, to search for competencies (knowledge systems projected to find specialists in various technical and scientific fields) and opportunities (by knowledge systems for finding opportunities for technical and scientific cooperation in various socio-economic sectors, from the demands made by companies).

14.4.3. *Information sources*

The Innovation Portal relies on sources of information from various sources. The Lattes System and the research groups' directory are the main sources, and they are the basis on which the portal is founded. In addition, other databases will be gradually formed as the instrument is being used: from recordings, from specialists, research centers, innovation and enterprise agents. Other sources of information can be added to the Innovation Portal system through interoperability and specific contracts.

The various areas of the Innovation Portal can open up a series of opportunities in the context of interoperability (interaction with other entities or units of information, integration, and information exchange, among others). In this sense, there are two contexts in which interoperability is provided on the Innovation Portal:

– entities wishing to make use of the Innovation Portal, for example, are for receiving information about experts or opportunities, among others, or, as the business environment provides information to the other Internet domains;

– consultation of correlative sources of information through the Innovation Portal i.e. users of the system can receive information from other databases.

The Innovation Portal is therefore considered as an important system, which is sufficient to support national strategic intelligence based on the university-industry-government relationship and for strengthening the ST&I system.

14.4.4. *Governance structure of the Innovation Portal for strategic intelligence*

The Innovation Portal is defined as an e-government instrument i.e. as an e-government service with specific objectives whose achievements depend on a "proprietary" body [WEI 06], but also on other actors, including the same hierarchical power, as in the case of the Ministry of Industrial Development and Foreign Trade (MDIC), through its direct action with the Brazilian productive sector.

In fact, considering the Innovation Portal as an instrument to support the university-industry-government interaction is perfectly correct. However, it cannot be reduced to a mere tool, as we have to observe its system of reference, and the advantage that made it an instrument with potential to help shape public policy related to its macro-environment. This, of course, depends upon the action of the Ministry of Science and Technology: it creates a space for a link (bridge) between supply and demand for knowledge in order to contribute the university-industry-government interaction.

As a result, there is a need for structuring a management that supports these functions, so as to enable the operationalization of the Innovation Portal, and which includes:

– regulation and management of relationships in the environment (internal) and macro-environment (external) of the Innovation Portal and that involve one or more responses to the interactions of its environment (internal and external relationships);

– operational and administrative management of the Innovation Portal to ensure the adequacy of infrastructure (hardware, software, human and other resources) necessary for its effective operation (exploitation);

– management of the content integrated in the Innovation Portal must ensure:

- the achievement of information flows necessary for the operation of its functionalities,

- the accuracy and integrity of data and information that will be introduced, as well as the results,

- the frequency of updating the content,

- the responsibility of each actor with respect to the introduced content and the use of information taken from the Innovation Portal (content);

– strategic management is responsible for the interaction between the strategic directive set in the macro-environment of the Innovation Portal and the objectives and strategies defined for the portal, taking into account:

- the diversity of interests to establish an interaction within the Portal,

- the e-government function, which imposes requirements (strategy).

In this context, management of the portal does not depend on the creation of new "entities", but the inclusion of specific duties for the preservation of the integrity and credibility of the Innovation Portal, both under the Ministry of Science and Technology and the delegations it has established, and the guarantee of these functions.

The strategy unit is equipped with features considering the functions that are associated with it, as it supports decision-making and strategic plan for the Innovation Portal. This unit is responsible for the strategic management of the Innovation Portal, where the model is inserted. Under the proposed model, the strategy unit's functions are:

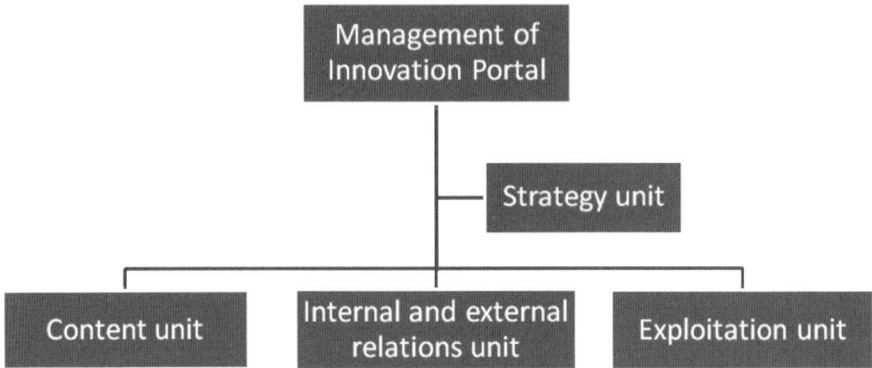

Figure 14.2. *Governance structure proposed for the Innovation Portal*

– to help in decision-making regarding the Innovation Portal, from the production of strategic information, watch, and the systematic evaluation of the Portal's macro-environment;

– to realize the strategic management of the Innovation Portal, as well as the strategic intelligence system of the Innovation Portal;

– to define institutional approaches for specific applications.

14.5. The strategic intelligence system of the Innovation Portal (SISIP): a tool for the Brazilian government

The strategic intelligence system of the Innovation Portal (SISIP) includes:

– a vision of the future represented by the strategic orientations of the Innovation Portal;

– the creation and distribution of intelligence to decision-maker;

– continuous analysis and evaluation of the environmental parameters related to the SISIP cycle.

However, it should be emphasized that the scope of application of the proposed SISIP is limited to the internal environment of the Innovation Portal, which means defining a monitoring system for the strategic management of the Portal; different from the Organization for Economic

Development Cooperation (OEDC) model, where the scope includes the monitoring of national innovation systems in member countries.

In this sense, the SISIP will act as a facilitating element for the strategic management of the Innovation Portal and the features of the strategic intelligence application distributed to its macro-environment will be:

– to identify trends and make recommendations for actions related to the Innovation Portal, as well as for its functionality;

– to provide strategic information for management and to groups of actors of the Innovation Portal responsible for the management;

– to allow the evaluation of strategic, tactical and operational orientations adopted for the Innovation Portal as well as modifications of these guidelines;

– to provide (measurable) elements for observing and analyzing the ergonomics of the Innovation Portal, in order to meet the requirements imposed by its presuppositions, such as e-government.

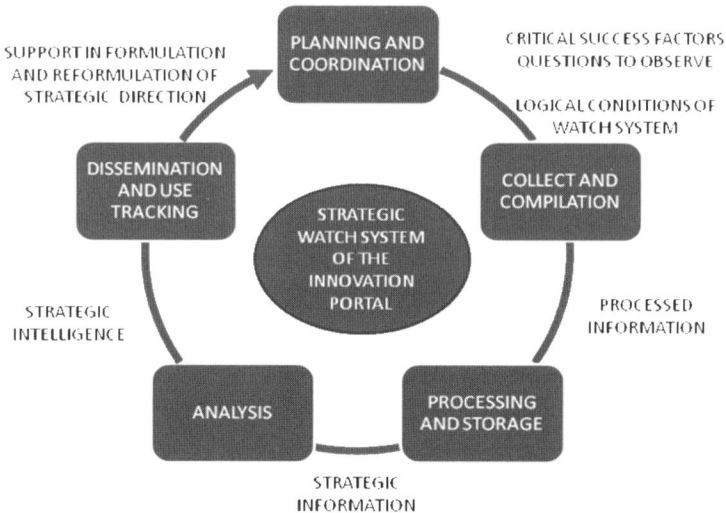

Figure 14.3. *Representation of the SISIP*

The application of the distributed strategic intelligence consists of an ordered sequence of actions, in a cyclical and systemic way, and attempts to generate strategic information for decision-making by more than one actor in this system.

The model proposed here, understood conceptually as an intelligence system, can also be defined as a research system of strategic information, to aid in decision-making on the Innovation Portal. Thus we will present below the system dynamics and stages of the intelligence cycle specified for the SISIP.

The dynamics of the SISIP, whose theoretical framework has been described, must consider not only specific characteristics, but also the following aspects: characteristics of decision-makers to be supplied with the results of SISIP and respective needs for information; identification of strategic information and critical success factors; collection and processing of information; analysis of processed information based on the agreed strategic frameworks; identification of responses to questions posed by the information needs through analysis – creation of intelligence; distribution of intelligence and watch and evaluation of products and processes in this system (or in this cycle).

The main implication of the SISIP is determined by the results of its dynamics, which seeks to help, understand and provide answers to questions posed by the strategic management of the Innovation Portal gradually as its stages (or rather the intelligence cycle established by the system) are developed.

14.5.1. *Planning and coordination stage*

The planning and coordination stage is intended to define the foundations of the intelligence system in order to perform the functions for which the system and the Innovation Portal were created. Based on what we have just stated, we were able to identify key elements of a strategic information system: the decision-makers and the extent of the Innovation Portal without ruling out the possibility of taking into account new elements.

14.5.2. *Decision-makers and scope of observation of SISIP*

Under the definitions of the system, we find the identification of key users i.e. those whom the system is intended for or those who need or have to share the results, as in any intelligence system. In this sense, the following are considered as decision-makers of SISIP or "receivers" of the results of the system: the management board of the Innovation Portal, decision-makers and policy-makers, representatives of stakeholders of the triple helix.

The board of management of the Innovation Portal is responsible for providing guidance to actions and activities related to the portal, and the assurance of full and effective operation: they are the main receivers of SISIP. Indeed, the system is useful to this board that responds to the full integration and representation of the Innovation Portal to the national innovation system and the Ministry of Science and Technology (MST). Decision-makers and policy-makers are concerned, after the board, as the most important receivers of the strategic intelligence system, since they deliberate actions and public policies relating to the macro-environment of the Innovation Portal and are also responsible for institutional approaches in the Brazilian system of innovation. They are represented by organizations or entities of the federal government having legal rights of action in the macro-environment of the Innovation Portal and relationships with the Ministry of Science and Technology (such as: FINEP, sector funds, the MDIC, ABDI and the Ministry of Science and Technology, among others). They are the actors of the Triple Helix (university-industry-government), who, through this representation, reinforce strategic information for their peers, after deliberation by the Board of Management of the Innovation Portal, according to the description of the dissemination and usage tracking stage of the SISIP cycle.

The observation scope of the Innovation Portal has set its macro-environment and can be specified as containing elements such as: ST&I environment, projects and programs related to e-government and information or knowledge society. We can thus define: time horizon, sectoring, sporadic consultations and thematic features.

The observations of SISIP must include a medium, and long term temporal horizon, considering 5 and 20 years respectively for this purpose; unless a specific request from the management of Portal Innovation, from the council for strategy or the Department of Science and Technology (the federal government actor represented in the macro-environment of the Innovation Portal may be different).

The SISIP could be applied to one or more specific sectors or themes, considering the creation of subsystems in the system itself. This implies considering and defining new elements which will determine the observation space of the new subsystem, as sectoral and thematic divisions have been provided for the usability of the Innovation Portal.

Sporadic consultations are considered for their temporal characteristic; they will be requested by the Strategic Management of Innovation Portal, which will define the parameters and characteristics.

14.5.3. *Critical success factors (CSF)*

The identification of critical success factors (CSF) was performed according to design and development specifications, with regard to the assumptions defined as the basis for the Innovation Portal. Meetings were conducted for specification, validation and discussion of spaces of the Innovation Portal, where the representatives of stakeholder groups also served as a reference for identifying critical success factors. SISIP takes into account, in part and to varying degrees, the CSFs defined throughout the contribution: for the national panorama and the Innovation Portal.

We consider that once FCSs are defined, it will be possible to establish a systematic of strategic watch for the Innovation Portal, included in the SISPI, given that the reference is the macro-environment of the Innovation Portal and that it takes into account issues relating to its main presuppositions: ST&I, university-industry-government relationships, e-government and information society.

The following are also considered as critical success factors for SISIP:

– the adequacy of the Innovation Portal to specifications of university-industry-government relationships for innovation. This CSF allows us to qualitatively and quantitatively review and analyze the university-industry-government relationships in a way to allow for the adequacy of the Innovation Portal functionality to the needs and demands of its macro-environment. Some of the aspects should be subjected to observation:

  - cooperation between scientific institutions and the business community,

  - attention focused on new growth areas by emphasizing collaboration and networking;

– compliance to funding policy and cooperation and innovation technology support policy. This means understanding the dynamics of innovation, the characteristics and idiosyncrasies of each player and of the innovation systems as a whole, so as to allow forecasting demand for an adequate interaction space, and thus ensuring the observation and adequacy of the functionalities of the Innovation Portal. Other aspects to be taken into consideration include:

- public funding of scientific research,

- efforts to reform the universities to grant them autonomy and emphasize their role as marketing research done through public funding,

- attention focused on issues related to the ST&I at the highest levels of government decision;

– adequacy of the Innovation Portal to the orientation and definitions established for federal e-government solutions. In this sense, observations on the following aspects should be included in the FCS:

- integrated projects at the federal level, particularly for the unification of the service delivery channels to citizens, to the civil society and the productive sector,

- sustainability and transparency of e-government initiatives at the federal level,

- systematic recognition of e-government initiative, to ensure standardization following the set guidelines,

- joint projects with the productive sector;

– maintenance of the management structure defined for the Innovation Portal, and provided with autonomy to manage the instrument and propose solutions to the macro-environment of the Portal. Compliance with this FCS implies taking into account aspects such as:

- qualified staff for new challenges in public administration,

- governance of e-government at the federal level;

– adequacy of the Innovation Portal to the conditions necessary for its usability and interoperability with its macro-environment, taking into account:

- inter-exploitability of systems being designed and those implemented,

- information security,

- infrastructure for handling and updating of e-government initiatives at the federal level.

Based on the above information, we can actually start the usability of the SISIP and, therefore, the strategic intelligence cycle set by the system. It should be emphasized that at the beginning of each cycle of intelligence, the basis, elements and conditions established here should be reviewed, so as to reflect changes and evolutions of the content of observation, and the involvement of the system. Also at the stage of planning and coordination, but now considering the strategic foundations established for the Innovation Portal, we can define the characteristics of information that will be needed, so that the FCS be observed, analyzed, etc. in order to orientate the information collection and processing stage. The phase described will be taken into account in the initial trial of SISIP, when the strategic framework of the portal would have been defined.

14.5.4. *Collection and compilation stage*

This stage entails identifying and validating sources of information to be used in the system and also establish the conditions for processing and aggregating data and information collected. This stage includes four well-defined actions, collecting available data and information; specific studies; processing of the collected material and evaluation of information sources.

For the SISIP, a large majority of information sources are available on the Internet, mainly on government or institution Websites which meet the interests of public or private as well as national or regional sectors, relating to partnership for innovation.

The following are considered as sources of information available for the SISP:

– the national and international Websites related to national innovation systems;

– national and international Websites dealing with relationships such as:

- industry-competitiveness,

- productive sector (which includes associations of the class)-university-government;

– databases of scientific papers on topics related to ST&I such as: innovation, public policy;

– triple helix;

– university-company relationship, etc.;

– databases of scientific papers on topics related to the knowledge society, the information economy, knowledge management and/or network management;

– the national and international sites dealing with e-government, e-governance, and e-development approaches; databases of scientific papers on topics related to e-government.

A second category of sources of information that will be considered by the SISIP will be sources specifically created for and by the system i.e. the specific studies carried out systematically and the sporadic studies. The followings are considered as sources of information created for, or by the system:

– the benchmarking study for the applications of e-government and e-governance; or rather for government solutions and e-governance;

– the benchmarking study on best practices in the university-industry-government relationship and Triple Helix for innovation (suggested frequency: 3 years);

– foresight study on the prospects of national innovation systems, impact on the Brazilian system, as well as on the characteristics and results of the dynamics of innovation.

The statistics are produced from the features and operations conducted in the context of the Innovation Portal. Studies are defined by specific demand from the analysis stage or from new questions or observations in response to the demand of SISIP, and are considered as sporadic studies. The results are provided by the studies and the statistics are incorporated into the SISIP with their information sources, and form an integral part of the established stream.

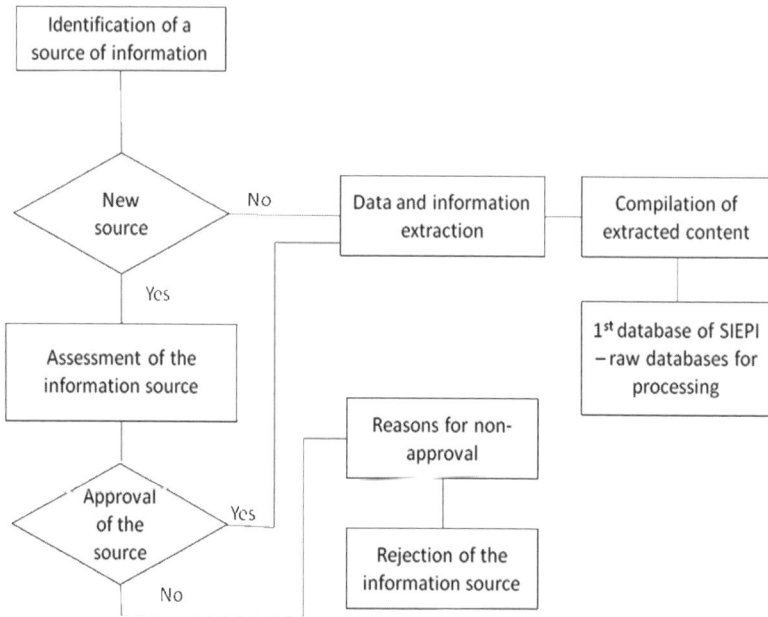

Figure 14.4. *The collection and compilation stage workflow*

The other activity of this stage relates to data and information processing from defined parameters, where it is also necessary to observe the predefined computer operational conditions, which may be related to the theme or the elements of strategic planning of the Innovation Portal. For processing data and information, two actions are paramount: the first relates to the collection and the second to the structuring of data and information: the preparation of material collected before treatment.

The definition of tools that have to be used is linked to computer solutions provided at the physical space, where strategic management of the Innovation Portal is carried out: this therefore introduces a different definition phase.

This step entails the evaluation of information sources, given that the SISIP must have a logical structure to absorb the changes that will certainly occur in the macro-environment of the Innovation Portal. In the medium and long term, these sources will be updated automatically and to do this, it is therefore necessary to incorporate a specific methodology defined by the activity of evaluating information sources.

14.5.5. *Processing and storage stage*

Since the system is an information system, it is obviously necessary to establish the physical and logical space where extracted and structured content will be stored. In the context of the SISIP, this compilation is given the name "processing space". The effective completion of this activity depends on definitions of the logic for processing (computer), and parameters are set by the strategic planning of the Innovation Portal, in order to identify consistencies and develop the necessary cross-analysis for produced content.

The information needs to be identified from the CSF of the Innovation Portal, and are the initial parameters to be considered, followed by questions defined from the strategic guidelines laid down in the short, medium and long-term or other questions identified in specific demands. The result of the processing activity is selected information from parameters and prefixed conditions. This information may be considered as "almost" strategic.

The storage activity includes keeping all the content included in the SISIP since the initial stage in a specific, logically and physically defined space, and therefore constitutes the formation of the second database which makes up the system and whose combination with other elements will form the basis of SISIP watch.

14.5.6. *Analysis stage*

The analysis stage is considered in the literature as the operating core of the intelligence system, since it reflects the incorporation of tacit knowledge and specialists' discussion on a specific topic, which seeks to build strategic intelligence as a result of the SISIP. At the end of the analysis process, we expect strategic intelligence would emerge as the result. The process of developing this highlights two levels of strategic definition: the first relating to the strategic direction and objectives set for the Innovation Portal, and the second relating to strategic frameworks adopted by the federal government for the macro-environment of the Innovation Portal. Both of them have been respectively established during and after the strategic planning of the portal. The analysis stage relies on the interaction with specialists or specialist groups, who may make the necessary analysis to SISIP in a more effective and more competent way. They could receive a portion of the intelligence produced by the system.

This stage of the SISIP also includes the administration of a basket of measurable elements (variables and/or indicators), which constitute a database on system statistics. Concerning databases, in addition to statistical database, information produced by the analysis process, such as those contained in reports, will be incorporated into previously structured databases, thus completing the processing space of the SISIP.

14.5.7. *Dissemination and usage tracking*

This step gives the system the ability to contribute effectively to decision-making related to the Innovation Portal, when in a structured way; it "gives" to previously defined "receivers" the results of the development of the intelligence cycle set in SISIP. In this sense, the place of the system should be detached from the management space of the Innovation Portal based on the relationships that will be established with the internal and external environments of the Innovation Portal. Thus, two "units" stand for the effectiveness of the SISIP: strategy unity, in which the physical and logical management of SISIP is integrated as well as its entire exploitation; the other important unit for the system is the internal and external relations unit, given that its purpose is to serve as the intermediary between the system actors and the macro-environment.

The relationship of the SISIP with the external environment of the strategy unit is done through the internal and external relations unit, so as to allow a single dialog "voice" and clearly defined strategies, which will be determined jointly. Another important point relates to the training of staff working for the exploitation and management of the unit, since they become "receivers" of tacit and explicit knowledge about the macro-environment with which they will eventually be in contact, because of their proximity to its players. The definition of the way this return will be made to SISIP must be defined at the same time as the dissemination strategies.

Dissemination of strategic information of the SISIP has more than one frequency, because of the "distributed" characteristics of system of intelligence and the fact that the main "customer" of the SISIP is a public official. In this case, the recognized main "client" is the management board of the Innovation Portal, which, once every six months, assesses the progress of actions, activities and relationships within the portal. Thus, the main report of SISIP called "Strategic Intelligence Report" (SIR) is produced twice in a year and responds to key questions from the CSFs of the Innovation Portal. An analytical report is prepared once a year. This report presents results from watch, and the applications of the results of SISIP in policies, programs, projects, activities or actions related to the macro-environment and the Innovation Portal. This annual report is called "Annual Intelligence Report" (AIR) of the Innovation Portal.

As previously defined, the management board of the Innovation Portal is comprised of members of stakeholder groups representing the Triple Helix (university-industry-government). Thus, at the same frequency as the SIR, specific reports per group of players will also be written. It is important to note that the strategic information resulting from the SISIP is not public. Although published at the same frequency as the SIR, group reports will be prepared after the review meeting of the board which will define, among other things, their scope and content.

Usage tracking of the results of SISIP will be made from a systematic watch, not computerized, to study, following the delivery of results, the possible impact of this result on "receptors" of the system, by observing their respective actions. Another approach, which will often be suggested to the strategy unit and unit of internal and external relations, is to provide a proactive activity to initiate debate on topics and/or specific results. Follow-up actions related to the observation does not depend on computerized systems internal to strategy unit, but those related to the storage of observed information should be incorporated into the SISIP in the processing space, integrating a critical mass of analysis object for the system. This process is called strategic watch. This process forms the essential basis of the strategic assessment.

14.6. Conclusion

If we structure the given context, one can deduce from it that the current landscape of the world and Brazil is and will be marked by constant changes in the dynamics of economic and social development of countries, organizations and societies; from where the increasing importance of the consolidation of national innovation systems would emerge, to encourage university-industry-government cooperation for innovation (the triple helix model), cooperation based on increased production and dissemination of information and knowledge through the use of modern media for communication and dissemination of results – virtual media.

Government institutions and governments must act in response to this new paradigm: the systemic vision, reflecting the extent of the action of decision-makers in the context of their responsibilities, without distancing themselves from the events of the globalized world or from the commitments made to the society. In this sense, we can consider that the function of information systems having the nature of strategic watch systems is essential for insertion and maintenance of countries in the 2.0 world, since, in addition to specific data and the possibility of integration with other systems or databases, they also incorporate specialized analysis for decision-making.

14.7. Bibliography

[AMA 04] AMARA N., LANDRY R., "Sources of information as determinants of novelty of innovation in manufacturing firms: Evidence from the 1999 statistics Canada innovation survey", *Technovation*, Otawa, 2004.

[BRA 04] BRASIL, Casa Civil, Lei 10.973, loi sur l'Innovação, 2004, available at: http://www.planalto.gov.br/ccivil_03/_Ato2004-2006/2004/Lei/L10.973.htm, consulted on: 12/01/2008.

[CAS 98] CASTELLS M., *La société en réseaux: l'ère de l'information*, Fayard, Paris, 1998.

[ETZ 98] ETZKOWITZ H., LEYDESDORFF L., "A Triple Helix of university-industry-government relations: Introduction", *Industry & Higher Education*, vol. 12, no. 4, p. 197-258, 1998.

[FOU 01] FOUNTAIN J., *Building the Virtual State: Information Technology and Institutional Change*, The Brookings Institution, New York, 2001.

[GUS 97] GUSMÃO R., *L'engagement français dans l'Europe de la recherche*, Economica, Paris, 1997.

[OCD 92] ORGANISATION DE COOPÉRATION POUR LE DÉVELOPPEMENT ÉCONOMIQUE (OCDE), Technological and the economy: The key relationship, Paris, 1992.

[PLA 09] PLATAFORME LATTES, Données de curriculums inscrits sur la Plataforme Lattes, available at: http://lattes.cnpq.br/#, consulted in March 2009.

[PLO 95] PLONSKI G.A., "Cooperação empresa-universidade na Íbero-América: estágio atual e perspectivas", *Revista de Administração*, vol. 30, no. 2, p. 65-74, April/June 1995.

[QUO 09] QUONIAM L., Résumé pour le Web 2.0, Course notes, Marseille, 2009.

[WEI 06] WEIKERSHEIMER D., CASTRO M.T., Advogados Associados, Parecer legal sobre o desenvolvimento, operação e gestão do Portal Inovação, p. 1-48, CGEE, Brasília, 2006.

Chapter 15

Regional Development 2.0

15.1. Introduction

Economic intelligence[1], which was used for over 10 years to increase the competitiveness of enterprises, can be used in a new field, for opening the door to many new exchanges and collaborations. While the end of World War II brought the development of East-West blocs and the Cold War, the fall of the Berlin Wall was the signal for a new era of global competition: globalization. We are entering a period of competition still unknown to date, where the rules of "the game" are no longer stable and can change quickly. At the same time, the development of information technology shortens "time and distance". This new era increases global instability, with competition taking place between societies, states, and regions. Currently, regionalization is a political way for reducing tensions and increasing the degree of personal freedom. In many cases, regionalization (this happens in all countries) is seen as an improvement and political conquest. But this regionalization in most of the cases is accompanied by a decrease in allocation of financial support by the central government, and therefore requires a major effort to create new resources. It is a major problem, since if these new resources are not created, the standard of living will decrease and political tensions will arise – political tensions that can lead to separatist movements or even terrorism. This will increase the instability of entire regions and will not

---

Chapter written by Henri Dou.

1 The term "economic intelligence" is strictly a French term. The term "competitive intelligence" is used for the same concept in other countries.

allow the realization of the desired economic development. For this reason, researches on methods and tools for creating a local value are increasingly being sought. In this framework, economic intelligence related to innovation, is one of the most powerful means to promote regional development. We will discuss economic intelligence in this chapter by highlighting methods, tools, brakes, and levers that lie along the course of its development.

15.2. Definition of Competitive Intelligence

We will not reconsider the various definitions of economic intelligence, but we will highlight some differences between the English-Speaking and European approach – in particular, the French approach. In the case of the United States, competitive intelligence developed largely by SCIP (Society of Competitive Intelligence Professionals) highlights the fact that the use of rational information facilitates decision-making in order to "dominate" the competition and provide the company with some competitive advantages. An extract of these definitions is as follows:

– "Systematic program to collect and analyze the information upon the activities of the competitors... in view to achieve the strategic goals of the company" (Larry Kahanner);

– analyze the information related to the competitors who are involved in the decision process of the company (Leonard Fuld);

– "Knowledge and forecast of the surrounding world – with a view to assist the decision of the company's CEO" (Jan Herring).

This emphasizes that the process of competitive intelligence is focused on information, company, and decision. In the case of Europe and especially in the case of French economic intelligence, the definition given by Alain Juillet (Senior Economic Intelligence Officer to the Prime Minister) is as follows: (excerpt) "develop a way of governance whose object is the control of strategic information which aims at the competitiveness and security of the (national) economy and (domestic) enterprises".

We see the emerging need to develop the national economy, and thus to focus economic intelligence toward national development, enterprise security, and in short, toward job creation. This is the vision that is found in Carayon's report, commissioned by Raffarin, the then French Prime Minister, and focused on economic intelligence, which is very explicit in the title of the report: "economic intelligence and social cohesion".

It is therefore within the context of implementing economic intelligence as a means for regional development which we will place the remaining part of this chapter. However, before addressing this, we need to define something that is almost essential in creating value (on the monetary side, exportation and job creation): the concept of innovation.

15.3. Innovation

Before addressing the fundamental mechanism of innovation, we note that innovation is currently perceived as the means to maintain and especially, to create competitive advantages. A whole series of reports highlights this aspect.

15.3.1. *Strong signals*

We will mention some of these signals:

– Palmisano Report: Innovate America (USA) "The U.S. Council on Competitiveness has unveiled a report entitled "Innovate America". Defining innovation as the "single most important factor in determining America's success through the 21st Century", the report clearly states that America's task in the next 25 years is to "optimize [the] entire society for innovation"". The Palmisano[2] report highlights the need for the US to innovate at all levels of society to determine their success in the 21st Century;

– Beffa Report: Renewing the French Industrial Policy (France), by enabling the R&D to be more innovative and competitive, for a new French industrialization;

– Renaissance II report (Canada) Canadian Creativeness and Innovation for the new millennium;

– Legislation to promote technology transfer from academia to industry (Law on TLO: 1998) (Japan);

– MITI (Ministry of International Trade and Industry) becomes the METI (Ministry of Economy, Trade and Industry) (Japan);

– The Council of Science & Technology becomes the Council for S&T Policy.

---

2 The name of the person (vice-president of IBM), who was responsible for its presentation.

15.3.2. *The mechanism of innovation*

Often, innovation is confused with invention. The mechanism of innovation is very different and we will present it in the light of the work done under the Interreg III[3] Programme of the European Community. During the presentation of this work, it was highlighted that, in the first step, the state will fund research and education, which will in turn contribute to the creation of knowledge and skills.

This is only a first step, and it should be noted that this step is mainly considered in the Latin world and especially in France as an end in itself. Yet this is not true, since the state cannot meet all the needs for laboratories and research development as well as for high education. It will therefore be necessary to build a second step which is innovation. The created skills and knowledge should be used to make money i.e. to create jobs and to export. This step characterizes innovation: knowing (from one's skills) to create new products and new activities which are likely to be well received by the market. This will lead to the development of new public-private partnerships and therefore the implementation of two research lines: one from Michael Porter [POR 81]: we innovate better in groups than alone[4], hence the concept of cluster development (competitiveness clusters in France), and the other – the Triple Helix developed by the Dutch research school several years ago [LEY 98].

Figure 15.1 shows the mechanism of value creation (i.e. return on investment).

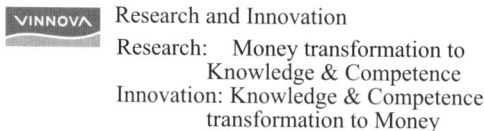

VINNOVA   Research and Innovation

Research: Money transformation to
                    Knowledge & Competence
Innovation: Knowledge & Competence
                    transformation to Money

Figure 15.1. *The basic mechanism of innovation based on the work developed in the European program Interreg III*

The concept of public private partnerships was developed from this vision of innovation and from the fact that we innovate better in group than

---

3 Centro Formativo Privinciale, Giuseppe Zanardelli, Azienda speciale de la provincia de Brescia, Interreg III C Brics-workshop – Aalborg, 13 February 2006, Dr Per Eriksson, Director General VINNOVA Swedish Governmental Agency for Innovation Systems.
4 Strategic intelligence and innovative networks. Stratinc, Interreg IIIC http://www. e-innovation.org/stratinc.

as a single entity (sharing ideas and skills). These partnerships which involve the state agencies (national or regional), research, education, and industry, form the core from which we can regionally create value.

We are not concerned about developing the set of actions that lead to the development of clusters, but we must know that in the context of developing countries, the creation of clusters is a fundamental step in regional development. It is in the cluster system that economic intelligence will find its place at the level of the overall governance and the cluster "roadmap". Figure 15.2 shows the structure of public-private partnerships (based on the works carried out within the European community).

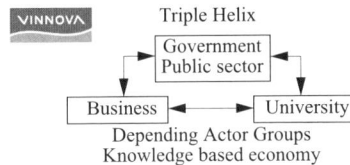

Figure 15.2. *Public-private partnerships, the triple helix*

A good example of this process was presented by Elias Zerhouni[5], director of the National Institutes of Health (NIH) in the United States: "The success of American scientific research depends on the existing implicit partnership between academic research, the government, and the industry. The research institutions are responsible to develop the scientific capital. The Government finances the best teams through a transparent system of selection. The industry holds the critical role to develop robust products intended for the public. This strategy is the key to American competitiveness and must be maintained".

15.3.3. *The support of innovation*

In some of the countries e.g. China, the focus is on "hard" technologies as well as on support technology. These two technologies, in line with teaching and research, will establish the conditions for value creation i.e. entrance into knowledge society. This way of exposing a problem is simple and has already been used successfully [DOU 06]. It is shown in Figure 15.3.

---

5 Introduced in December 2006 during the congress organized by the American Society of Hematology. Cited in "*quel modèle pour la recherche publique française?*" les échos, January 10, 2007, Alain Perez.

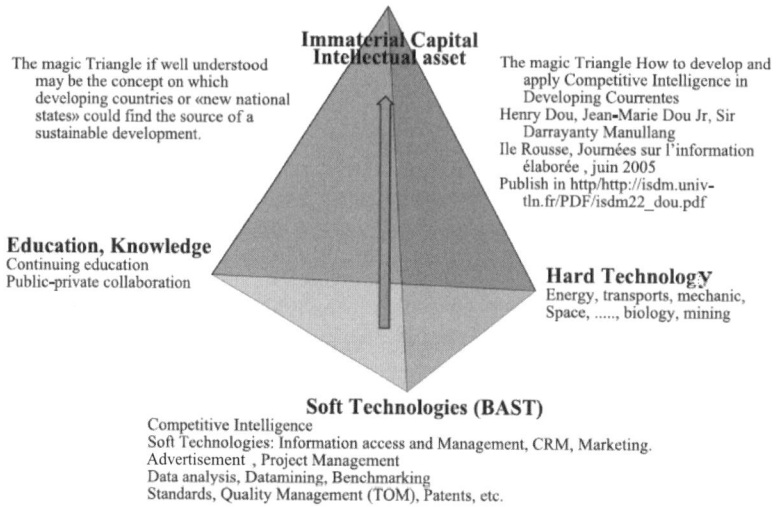

The magic Triangle if well understood may be the concept on which developing countries or «new national states» could find the source of a sustainable development.

Immaterial Capital
Intellectual asset

The magic Triangle How to develop and apply Competitive Intelligence in Developing Courrentes
Henry Dou, Jean-Marie Dou Jr, Sir Darrayanty Manullang
Ile Rousse, Journées sur l'information élaborée , juin 2005
Publish in http/http://isdm.univ-tln.fr/PDF/isdm22_dou.pdf

Education, Knowledge
Continuing education
Public-private collaboration

Hard Technology
Energy, transports, mechanic, Space, ..., biology, mining

Soft Technologies (BAST)
Competitive Intelligence
Soft Technologies: Information access and Management, CRM, Marketing.
Advertisement , Project Management
Data analysis, Datamining, Benchmarking
Standards, Quality Management (TOM), Patents, etc.

Figure 15.3. *The magic triangle for developing intellectual capital*

At the base of the triangle, we place the "hard" technology, soft technology, as well as education and knowledge creation for action. The term "hard technology" has been well understood, whereas, the term "soft technology" is less perceived. It includes all support technologies This will enable to the R&D decesion makers to pass the results obtained from the hard technologies to the market context, and make them sustainable for the society. In this context, economic intelligence has a prominent place. Currently, there is a strong movement in this direction especially at the Academy of Soft Technology Beijing (BAST) [ZHO 02] in China [DOU 02].

The combination of these three areas leads to the creation of intellectual and intangible capital of companies, organizations, institutions, and regions. We therefore enter into the knowledge society, with the creation of actionable knowledge [DOU 04] to ensure that sustainable development [BAR 01] is a prime objective of economic intelligence. Therefore, another pillar (that of knowledge) is added to the three classical pillars that led to the development of the industrial age. It is no longer possible to reason as in the past; we must overcome the constraints of our lifestyle, our education, and economic rules and policies. We must think "out of the box".

15.4. An introductory example: South Korea

Often, the Japanese development is considered as an example by highlighting the importance of analyzing formal and informal information and the introduction of strategic aspects in business development system. However, Japan [ISH 06], though partially destroyed during the Second World War, had (before the war) a sufficient level of technology to quickly rebuild its industry.

We would prefer to consider an example of South Korea, which in 40 years has moved from a subsistence agrarian economy to an economy of developed country, with the creation of high value-added products and brands like Samsung, Hyundai, LG, etc. A few years ago, during a South Korean visit to Morocco, a presentation was made by the South Koreans. This presentation highlighted a number of steps and strategies that led to the achievement of the current development. Figure 15.4 shows the importance of the development of South Korea (which before the Second World War had virtually no industry, but just basic subsistence agriculture) and the overall result.

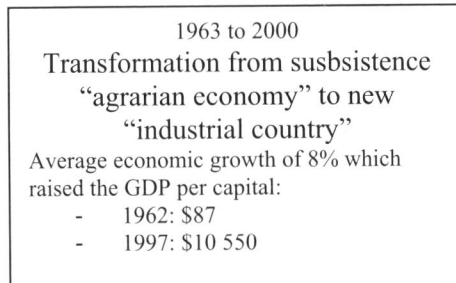

1963 to 2000

Transformation from susbsistence "agrarian economy" to new "industrial country"

Average economic growth of 8% which raised the GDP per capital:
- 1962: $87
- 1997: $10 550

Figure 15.4. *The rise of South Korea*

Such economic development activities cannot be improvised, and in the following presentations, the South Koreans highlighted:

– the need for policy continuity over a long term i.e. a consensus on development priorities and actions to implement;

– the creation of a national intellectual capital i.e. the ability to have elites who are likely to understand and integrate the technologies and scientific developments from abroad;

– development of a national information system;

– creative imitation which we may call incremental innovation i.e. the establishment of successive steps to achieve a fixed objective;

– the mobility of scientists, allowing them to work in foreign laboratories and acquire the required skills, through well-oriented programs;

– management of foreign direct investment i.e. the organization of technology transfer (e.g. in the context of cooperation agreements);

– the development of private industrial conglomerates and the formation of a strong national industrial base.

The country's transformation evolved through this policy, which was followed almost without interruption for more than 40 years. By considering the various aspects of this policy in detail, we can note that without citing clusters or economic intelligence, methods and tools used to develop clusters and to innovate were applied by the South Koreans.

Before highlighting other examples, it is important to have an accurate view of what incremental innovation is (see Figure 15.5).

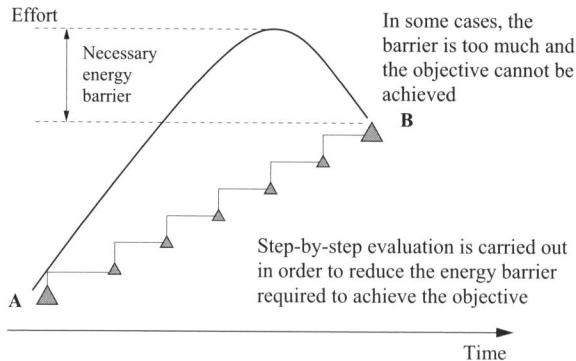

Figure 15.5. *Incremental innovation*

In the developing countries, leaders would often like to catch up very quickly, or they have technological ambitions to develop a weapons program, transportation program, heavy industry program, agriculture programs, etc. These programs cannot be carried out quickly and would require a continuity in the action and development of scientific and technological resources, as well as of information systems, in order to succeed. In addition, it is often necessary to change attitudes at the same time, and this cannot be done quickly. The same method of step-by-step evolution is also applicable at the social level. To achieve these objectives, it is essential to develop programs that consider all the scientific, technological, financial, and social aspects. This does not happen without the establishment of a general policy, e.g. economic intelligence, which, by focusing energies on strategic objectives and by ensuring continuity in time, will achieve the expected goals.

In this way, the industry cluster policy was developed in South Korea [CHA 09]. Different types of clusters were developed. There are four main types of clusters[6]: clusters with the Academy (research and education) as the leader, specialized cluster from a local community, cluster with large companies as the leader, and Silicon Valley type cluster. The first cluster Daedeok was focused on research and academy. Currently, various efforts are being made to transform it into a Silicon Valley-type cluster: "Daedeok 21 Century".

The different locations of clusters in South Korea are represented in Figure 15.6.

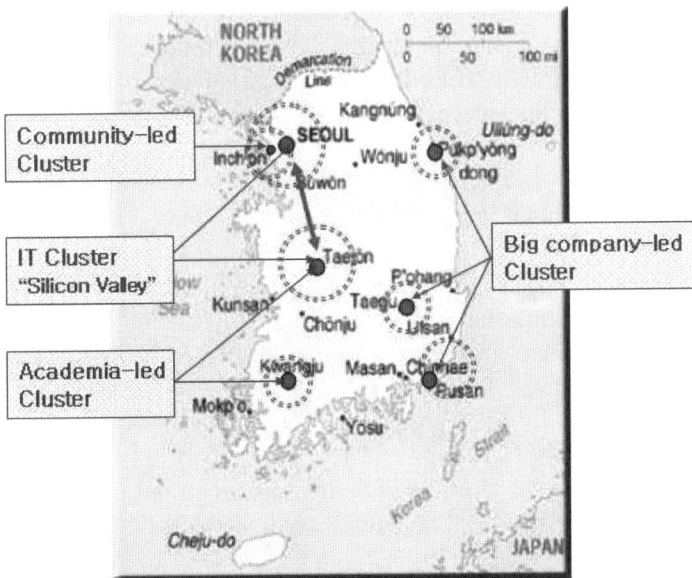

Figure 15.6. *The location of different clusters in South Korea and their dominance*

6 http://www1.american.edu/initeb/hp2566a/IT%20Geographic/new_page_3.htm.

15.5. Other examples of cluster development

The concern is not about developing a comprehensive study of the use of economic intelligence in the development of clusters, but to show the usefulness of this approach through different examples.

15.5.1. *Malaysia*

Malaysia has transformed from an agricultural to an industrial economy in 20 years [KHA 01]. The process that allowed the transformation is quite different from that used in other developing countries. The industrialization process of Malaysia was initially based on the political will to create local products which could substitute for import products.

To encourage such substitution industries, the government has directly and indirectly subsidized the establishment of new factories and has protected the domestic market.

To facilitate the integration of Malaysia into the new international division of labor, measures such as the introduction and establishment of free trade areas have been initiated. The government's ability to attract foreign investors who are seeking to reduce production costs in order to be more competitive in the international market, is a reason for the success of the industrialization.

This industrialization was achieved in two stages. Initially, parts or imported products parts were made locally. Local manufacturing of these parts were made possible by creating joint ventures, for example with Japanese firms. Such joint ventures were initiated to improve product reliability, and enhance and recognize the ability of Malaysia to produce advanced industrial components.

In a second stage (period 1996-2000), they developed research capabilities necessary to, from basis technology acquired in the first step, to realize endogenous creation of new products[7] [MAH 98].

---

7 Ministry of International Trade and Industry, Second Industrial Master Plan 1996-2005, Executive, Summary, Kayzn-Klerr (M), Sdn Bhd, KL, p. 12, 1996.

Thus, from this policy, international brands like Samsung have developed a range of products exported to international markets. At the same time, a local industry with a strong economic impact has been developed with the establishment of a range of Proton cars[8]. These are beginning to be exported to the United Kingdom, South Africa, Australia, and the Middle East. This forms the main exports. In 2006, approximately 14,800 Proton cars were exported.

But Malaysia did not confine itself strictly to industrial cluster development; it has also developed clusters based on natural resources e.g. palm oil and biofuels.

It is currently one of the largest producers of palm oil[9] in the world and even if the price of palm oil has fallen due to the current crisis, this cluster makes a significant contribution to the Malaysian economy.

15.5.2. *Thailand*

Thailand was able to develop a quality information system through patent analysis, for example. At the same time, it developed an aggressive policy of creating clusters based on its natural advantages for competitive advantages. This policy of cluster development was supported by the United States via the USAID agency.

KI.ASIA site provides access to many resources in the field of cluster development. We can cite among others: Cluster World Book, competitiveness and clusters for SME, competitiveness versus clusters, improving one's standard of living from the clusters; wear and clusters; Thai competitiveness from five clusters, the process of "clustering", Porter's competitive strategy, and Porter's five forces for the industry. This information is accessible via the "Downloading Center" of KI.ASIA[10] (free registration).

The choice of these clusters has been achieved as a result of the cartography of the local potentialities at both hard and soft levels. This led to the establishment of a niche policy which is shown in Figure 15.7.

---

8 http://en.wikipedia.org/wiki/Proton_(carmaker).
9 Palm Oil Research Institute of Malaysia (PORIM).
10 http://www.kiasia.org/EN/Myresume.asp?.

Vision communicates feasible and ambitious goal	Vision is removed from cluster's current challenges	Vision fails to set the right direction for the cluster
Automotive: Detroit of Asia	Fashion: Asia Tropical Fashion	Software: World Graphic Design Center
Tourism: Asia Tourism Capital	Food: Kitchen of the World	

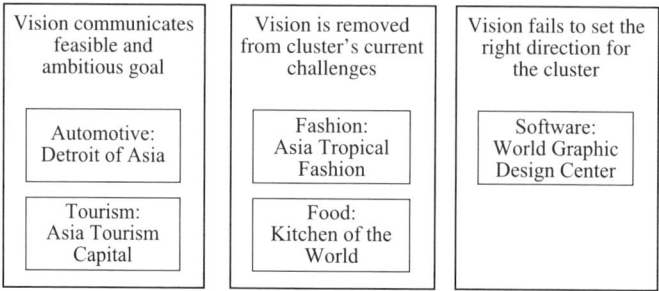

Figure 15.7. *Niche policy in the development of clusters (other clusters based on skills and international demand may be added to these areas)*

For example, the outlines of the tourism cluster are shown in Figure 15.8. In cluster development, all stakeholders must be involved and this will lead to more rapid dissemination of information, synergy between different skills, and rapid development of products.

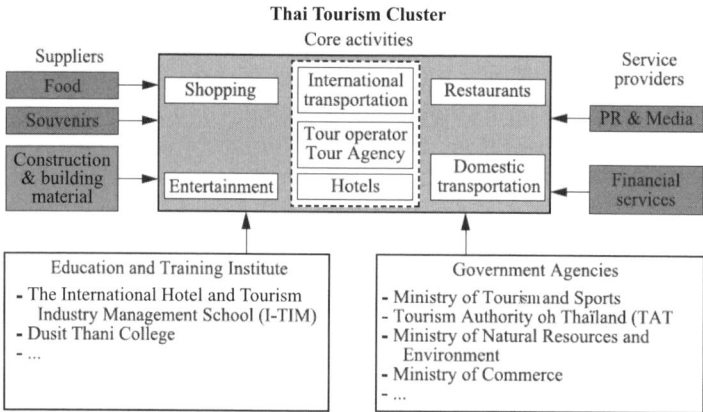

Figure 15.8. *Stakeholders in the tourism cluster*

The formation of a cluster usually involves a general analysis of all available data on the area considered (localized cluster), whether at the level of raw materials, industry, skills, release, transport, etc.

In the study conducted, which concerns the zone of Fos in the Provence-Alpes-Cote d'Azur, a complete cartography of the zone, including environmental and sociological aspects, was performed. It is a good example demonstrating the feasibility studies to be made before the establishment of a cluster [MOI 08].

15.6. The "pre-clustering" in developing countries

The establishment of clusters is generally performed when an industrial potential is available along with a set of skills, and also a local political will. But in some developing countries or in certain regions of these countries, it is not possible to directly develop clusters, in the industrial sense of the word. It will therefore be necessary to go through successive stages, and the first steps will be to start from an existing natural resource and then create community groups to begin work on this resource using relatively less advanced and locally available technology. One way to approach this goal can be through cottage industry, for example, or by the allocation of financial resources such as microcredit or microfinance programs.

This approach of "precluster" can only develop if there is a local political will to "redistribute the cards" i.e. not to remain in a status quo of using the natural resources as a basic material sold as such, and not producing sufficient added value (except for a few) to improve the general well-being.

Another situation encountered for example, in India, concerning the culture of spices is the production volume and quality. At the village level, there is a need to organize the harvest of spices and their conservation methods to ensure a quality production that will enable dialog with the distributors. Thus, the cottages industries will evolve into cooperatives, which can then be grouped into clusters to introduce changes based on technological development.

But to achieve this goal, it is necessary to develop, regionally, some capacity in R&D (Science and Technology) and in information to acquire the necessary basis from which to evolve the cooperatives toward a "precluster". The skills can be achieved using means that are currently available. Figure 15.9 shows how knowledge can be acquired [WAT 08].

National Technological Learning

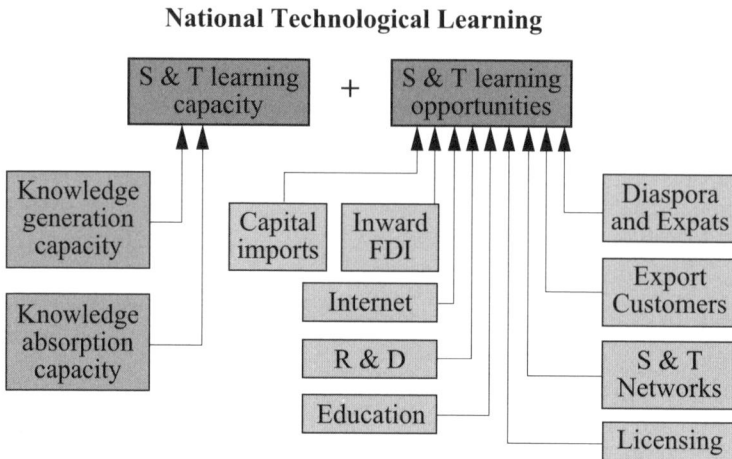

Figure 15.9. *Means of acquiring technological knowledge*

15.6.1. *Example of Sri Lanka*[11]

Spice production is carried out in Sri Lanka by smallholders who have less than 1 hectare of land to exploit. The main spices that are exported are: cinnamon, pepper, cloves, cardamom, nutmeg, and mace. Cinnamon is about 50% of the total spice exports, representing about 1% of total exports and 7.6% of total agricultural exports. There are currently over 400,000 small farmers who harvest 70% of the total production of spices. Currently, small farmers sell their produce to middlemen who then resell the products to processors to obtain powdered spices, sachets, etc. We have cited a classic example of a cluster; this will avoid intermediaries and will better compensate producers.

In 2001, Sri Lanka spice cluster headed by the "spice council" was created with the assistance of USAID (United States Agency for Development). The leaders will ensure the governance of the cluster and ensure compliance with the "roadmap" of the latter. This cluster includes all stakeholders in the spice industry, from production to processing. Among the whole participants, we find configuration of public/private partnerships – foundation for the development of clusters.

---

11 The Spice Council c/o The Agribusiness Council.

The spice council missions are as follows:

– facilitate and implement competitive strategies to achieve the industrial vision by 2010;

– act as an apex for the spice industry in Sri Lanka, and meet the aspirations of all stakeholders by unifying their efforts to achieve the industrial vision;

– to be very innovative and create a value oriented toward customers' needs;

– to promote the position of Sri Lanka in the world markets and contribute to the national economic growth.

The followings can be noted among the strategic directions of the cluster: quality improvement, quantity improvement, and consistency in the supply of spices. This initiative seeks to strengthen ties between exporters, processors, and small organized farmers to stimulate the development of the spice plantations and improve agronomic services for farmers.

15.6.2. *Example of Indonesia*

The same situation exists in Indonesia, with regard to various natural resources such as coconuts, cloves, nutmeg, different types of nuts, vanilla, cane, palm oil, etc. In most cases, we have only one harvest and direct resale without further processing, while in other cases such as coconuts, where industrial production is most advanced, only low added value products are produced. Yet, certainly there are scattered skills in these areas in the country, but a comprehensive information system and a political will to change the existing situation are lacking i.e. to encourage the development of various higher value added industrial products, such as cosmetics, essential oils, etc. without waiting for the arrival of investors.

In the case of natural products which can be used in cosmetics, the major problems that were faced concern the quality level, starting from the harvesting of product to their transformation. There is no real control or quality label to improve Indonesia's position in these areas.

Cluster development in these areas would be through the establishment of a national or regional information system that would allow knowing what is produced and protected in these areas, alongside opening toward the companies that are in the sector and toward the international market prices, etc. It is also notable that this should be accompanied by capacity building at the local university, even if it (the university) is inefficient in terms of research. In fact, skills can be learned in different ways, as discussed above.

15.7. Conclusion

We have seen in this chapter that regional development could be assisted by an economic intelligence initiative, an initiative that would lead to the development of clusters or "preclusters" and a new public-private partnerships structure (for developing countries where universities-research-industry relationships remain weak and poorly organized).

According to the initial industrial levels, the procedure to be followed may be different, but it must still go through the development of a quality information system oriented toward natural resources to be valued. At the same time, either at the level of laboratory and research, or by endogenous or exogenous acquisition, skills must either be created or strengthened. The foundation for development can be laid from this basis and a strong political will. It is also obvious that in many cases it is necessary, with the desired technological development, to change attitudes. This can only be done through socialization of projects and a kind of regional or national consensus oriented toward development.

These local initiatives have also the merit of structuring projects, in real time, and therefore present a range of possibilities, potentially attractive to investors. Whether at the level of natural resources, small industry, handicrafts, or tourism, the economic intelligence-cluster-public-private partnerships "mechanics" remain the same.

We should also note that often these initiatives should be "catalyzed" by agencies such as USAID (in the case of Sri Lanka and Thailand, for example), because these value creation systems are new and policy-makers are not familiar with them. Similarly, at the university level, those in charge will be reluctant to develop training initiatives and extension of economic intelligence, seeing this new discipline as a competitor to the existing areas of education such as Physics, Chemistry, Marketing, etc.

At the level of developed countries [MAS 06], the use of economic intelligence as a vehicle for economic cooperation [DOU 05] works well, whether in the context of the French-speaking countries (for France), or through bilateral actions[12] (Chile, Spain, Portugal, Brazil, etc.). This proves that the use of economic intelligence as a vehicle for regional development is a method that is now widely applied.

To conclude this chapter, we will quote Charles Gave [GAV 03]: "Toynbe in his famous book *L'histoire* speculates that the role of elites in a country is to meet the historic challenges that are thrown to the nation. If these elites are unable to meet the challenge, the same challenge will be repeated until it results in: at best, a change of elites, usually the extirpation, and at worst, a collapse of civilization. The challenge now is to get out of economic and especially intellectual stagnation in which we're stuck". This was published in 2003, well before the current crisis, and we must strongly think on this.

15.8. Bibliography

[BAR 01] BARONI DE CARVALHO R., AROUJO TAVARES FERREIRA M., "Using information technology to support knowledge conversion processes", *Information Research*, vol. 7, no. 1, October 2001, available at http://Information R.net/ir/7-1/paper118.html.

[CHA 09] CHANG H., CHI Y.J., "How to create industry clusters: A comparison of industry cluster creation process in South Korea and the United States", *The Annual Meeting of the Midwest Political Science Association 67th Annual National Conference*, The Palmer House Hilton, Chicago, USA, 4th August 2009, available at http://www.allacademic.com/meta/p360491_index.html.

[DOU 02] DOU H., ZHOUYING J., "Passer de la représentation du présent à la vision prospective du futur", *Technology Foresight, Humanisme et Entreprise*, December 2002.

[DOU 04] DOU H., DOU J.M. JR, "The processes of building knowledge. The case of smes and distance learning", *ISDM Information Science for Decision Making*, vol. 17, no. 174, June 2004, entire text available at http://isdm.univ-tln.fr/articles/num_archives.htm.

---

12 Most of the remarkable initiatives in this area (with respect to France) are available at http://www.ciworldwide.org.

[DOU 05] DOU H., DOU J.M. JR, MANULLANG D.S., "The magic triangle. How to develop and apply competitive intelligence in developing countries", *Journées sur l'Information Elaborée*, Ile Rousse, June 2005.

[DOU 06a] DOU H., "Competitive intelligence a new backgroud", *Conference*, Institute of Scientific and Technical Information of Shanghai, China, 14 November 2006, entire text available at http://www.ciworldwide.org.

[DOU 06b] DOU H., "Competitive intelligence accelerator of cooperation", *Intelligencia Economica defensa y seguridad – Que desafios por el siglo XXI*, Ecole Militaire, Santiago du Chili, p. 21-22 November 2006, entire text available at http://www.ciworldwide.org.

[GAV 03] GAVE C., *Des lions menés par des ânes*, Robert Laffont, Paris, 2003.

[ISH 06] ISHINYA O., Industrial Cluster Policy in Japan, Regional Economic and Industrial Policy Group METI, Government of Japan, 22 April 2006.

[KHA 01] KHAIRUL A.A., ABD R., SUGIYAMA K., WATANABE M., MAHATIR M., Knowledge Conversion Strategies for Malaysia Industrial Clusters, Working paper, University Malaysia Sarawak, 2001, available at http://www.idemployee.id.tue.nl/g.w.m.rauterberg/conferences/CD_doNotOpen/ADC/final_paper/170.pdf.

[LEY 98] LEYDESDORFF L., ETZKOWITZ H., "The triple helix as a model for innovation studies (Conference Report)", *Science & Public Policy*, vol. 25, no. 3, p. 195-203, 1998.

[MAH 98] MAHATHIR M., *Multimedia Super Corridor*, p. 20-21, Pelanduk Publications, Kuala Lumpur, 1998.

[MAS 06] MASSARI G., DOU H., QUONIAM L., DA SILVA H., "Cicera, Ensino e Pesquisa no campo da Inteligência Competitiva no Brasil e a Cooperação Franco-Brasileira", *Puzzle, Revista Hispana de la Inteligencia Competitiva*, vol. 6, no. 23, p. 12-19, August-October 2006, available at http://www.revistapuz zle.com\puzzle_sum_23.htm.

[MOI 08] MOINE H., CLERC P., DOU H., "Intelligence Economique et Développement territorial; La zone industrialo-portuaire de Fos", *Oriental.ma*, no. 4, p. 25-29, December 2008.

[POR 81] PORTER M., *The Competitive Advantage of Nations*, Free Press, New York, 1981.

[WAT 08] WATKINS A., "Building STI capacity for sustainable and poverty reduction", *The AAAS Meetings, World Bank*, Boston, 17 February 2008.

[ZHO 02] ZHOUYING JIN, Driving forces for sustainable development, AI&society vol. 16, no. 1&2, 2002.

Chapter 16

Government Strategies of Territorial Intelligence 2.0: Support to SMEs-TPE

16.1. Introduction

Public policies reflect different development models adopted by states. Under the influence of economists such as Douglass North (1990–1994)[1], Amartya Sen (1997, 1999, 2000)[2], and Joseph Stiglitz (2002)[3], the theme "inequality", previously a concern in social policy, is fully incorporated into economic strategies.

Based on this orientation, nations in their development efforts seek to eliminate or at least alleviate social inequalities. It is believed that poverty leads to reduced freedom, since it prevents the ability to act (to perform a function, work, or undertake and conduct business). Reducing poverty, therefore, means giving positive freedom to individuals and institutions. What becomes really important is their capacity of "doing and being". Positive freedom means allowing individuals and institutions to operate in the society, since it becomes part of and, at the same time, an instrument of development; their instrumental role extends beyond economic production since it leads to social development (Sen: 1997, 1999, 2000).

Chapter written by Kira Tarapanoff, José Rincon Ferreira, and Lillian Alvares.
1 Nobel Prize, 1993.
2 Nobel Prize, 1998.
3 Nobel Prize, 2001.

Such an orientation has an influence on the global public policy. We will consider, in particular, the establishment of information and business telecenters[4] (the TIN network in Brazil) for micro and small enterprises (MSEs).

The idea of information and business telecenters was adequate to the Brazilian reality. In fact, our data show that the main barrier faced by the MSEs in Brazil is the difficulty in developing the skills necessary to access information and communication technologies (ICT) and their usage. Lack of knowledge and excellence in ICT usage prevents them from leveraging the dynamics of communication and market transactions dominated by the digital economy.

From this hypothesis, MSEs (potential generator of employment and increased income) may make intensive use of ICT to increase their competitiveness, thereby enabling the expansion of their businesses, faster communication, cheaper access to useful information, more agility in buying and selling, expansion of their markets, and reduction in their operational costs. Therefore, electronic commerce is the cornerstone of the new economy since it provides information and resources necessary for people and businesses to operate in the new modes of business and commerce.

The TIN network is a space designed primarily for the economic and social achievement in MSEs, and is different from other social or educational computer literacy projects. In addition, the network also ensures the development of its capabilities, so as to emerge more competitive through the conveyed content.

The TIN network goes beyond digital and social inclusion. It particularly enables "intelligent" development of businesses and of communities.

This chapter shows that the success of the TIN (Information and business telecenters) network results from the application of the 2.0 concept.

16.2. Elements of the 2.0 concept applied to the TIN network

The establishment of information and business telecenters program has provided Brazilian society with the chance to implement the 2.0 concept. Its constitution necessarily includes the presence of five different cases.

---

4 Telecenters were set up in 1985 in the Nordic countries as digital inclusion programs and access to technology for the general population, called community teleservice centers (STSC).

The first three are structural. We must understand the project in the context of cooperation and participation, and focus on the social concerns, and seek the application level of the 2.0 concept. The immediate result of these three situations is the outlining of a perpetual beta context – leading to the fourth case – where everything is continually improving. This context, based on the users' responses on quality and treatment of services, allows us to incorporate the users' reactions for continual of the product (meaning that we will never get to a stage of finished product but that of a product continual improvement).

The fifth case is the central goal of what has been said above, the basic need to target users of the TIN network, where the users are mainly localized in specific contexts instead of being in a mass context, also known as "Long Tail[5]". Figure 16.1 illustrates the 2.0 context.

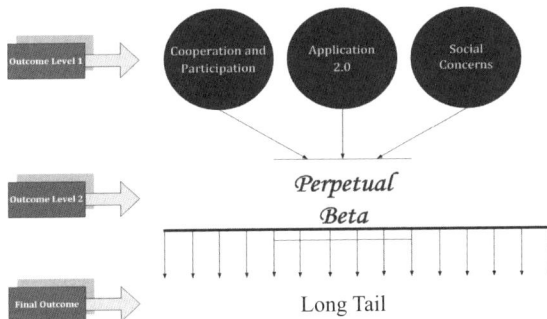

Figure 16.1. *Schematic diagram of the 2.0 concept*

16.2.1. *Context of cooperation and participation in infrastructure development*

Since its origin, the TIN network project is the result of a collective effort. The idea of setting up the project was defined under the "Permanent Forum of Micro and Small Enterprises" and its implementation relies on the partnership of around hundred entities.

The first stage of the project was to install 866 TIN, for forming a national network in all the regions of Brazil. We hope that over the next few years at least one telecenter will be installed in each Brazilian municipality

---

5 Long tail global micromarket.

through partnerships. At present, we can approximately count 5,000 authorized units, out of which 3,000 are being installed, and about 1,500 are fully operative.

The TIN project is, in fact, composed of a network of networks. Their management, planning, supervision, and operationalization are complex activities which involve different public and private institutions. We must, therefore, pay particular attention to the construction of an organizational model which reflects the diversity of components, the role to be played by institutions and collaborating participants, and the role of the telecenter itself. In addition, the structure must be robust enough to meet its own needs by becoming independent of government action.

The initial proposal, which should be improved with the proper functioning of the network and experience sharing during appropriate discussion forum, foresees various actions to be taken.

16.2.1.1. *Coordination and mobilization units*

Coordination and mobilization units are the institutions that are saddled with the responsibility of coordinating and supervising the setting up and operation of TIN network units. In addition, these units have to mobilize and motivate the presentation of proposals for the establishment of telecenters and the bonding of other collaborating institutions of the network. They work under the close guidance and supervision of the central management and supervision system.

16.2.1.2. *Content providers centers*

Content providers centers consist of production institutions which are charged with maintaining and disseminating contents, where the contents are available for access on the network, for consultation. The contents, if required, can be sold to the MSEs who are served by the telecenters. They operate according to the standards established by the TIN network, and are oriented based on satisfaction surveys conducted among the clients served and on requests for products and information.

16.2.1.3. *Sponsoring and supporting institutions*

Sponsoring and supporting institutions consist of public and private entities, where these entities provide physical, material, and financial resources for supporting or sponsoring the installation and operating one or more telecenters.

16.2.1.4. *Strategic partnerships*

Strategic partnerships are important national and international entities which offer their support in the political circles by obtaining funds, negotiating and releasing government subsidies, as well as implementing of agreements and conventions for the implementation, operation, and development of the TIN network.

16.2.1.5. *Coordination and management committee*

Coordination and management committee concerns the forum of deliberating college of TIN network. The Coordination and management committee consists of representatives from the installed telecenters, the coordination and mobilization units, the secretariat of industrial technology of the Brazilian Ministry of Industry and Foreign Trade Development, the contents provider centers, the institutions that sponsor and subsidize, the strategic partnerships, the users, the director of the central management and supervision unit, and the coordinator of information forum for the MSEs. This forum is responsible to guide, plan, evaluate, and define the policy and strategic lines of the entire TIN network. The forum deliberates, directs, and approves the strategic, marketing, distribution, operational, and implementation plan of TIN network resources

16.2.1.6. *Central management and supervision unit*

The central management and supervision unit is responsible for the management, evaluation, supervision, and execution of implementation, operation, and expansion of TIN, as per guidelines and plans approved by the coordinating committee and management.

The OSCIP[6] (Social Organization of Public Interest) of business and information telecenters (ATN[7]) will gradually assume this responsibility, so that it can replace the MDIC during operational actions, such as service provider, and by providing the necessary resources for the project.

---

6 OSCIP: Denomination created by the Brazilian Ministry of Justice to designate a type of non-governmental organization (NGO). It is regulated by Law 9790 dated 1999.

7 ATN: Created in March 2006, ATN is dedicated to the development and sustainability of the telecenters of business and information.

Gradually, the OSCIP would seek to ensure that the entire network becomes autonomous and is independent of government action, so as to prolong the existence of the existing created structure.

16.2.1.7. *Users*

The user category includes all the registered users and those who are serviced by a telecenter, whether physical or legal persons, or artisans, or entrepreneurs, or heads of MSEs. Users can make use of services by using the physical facilities of a telecenter or connecting to the site of the telecenter from any Internet base.

16.2.1.8. *Providers of data communication services*

The establishment of the TIN network is based on the usage of the existing communication infrastructure, in different regions and localities where these channels have been estimated based on flow of information and data, as well as the Internet connection which is subject to the available broadband.

16.2.2. *Social movements for the inclusion of other segments*

Telecenters, with their variety of modalities, including those installed in the poorest regions of Brazil, are definitely part of the country's development agenda. They promote the reduction of social inequalities and facilitate access to information for the socially excluded. The TIN network of telecenters is located in poor neighbourhoods of large urban peripheries, as well as in more remote areas and rural communities in various regions of Brazil such as Amazonia.

Brazil has more than 5,000 telecenters. These telecenters are supported by the federal government, public enterprises, and other organizations. They allow thousands of people who do not have the resources to acquire a personal computer or have Internet access in their homes, so as to have an opening to the world of knowledge and opportunities.

The TIN network is more involved in the social sector by way of introducing the formation, as well as consolidating the food vigilance and local development consortium. This consortium is involved in systematizing and disseminating the information on priority thematic areas to strengthen human and social capital, public management capacity, with emphasis on the work and the income-generating capital.

We can differentiate between the acquisitions for family farming and the establishment of local socio-productive arrangements, in this consortium, with the objective of solidarity in the economics of microfinance, handicrafts, and electronic commerce.

The TIN network has been established in coordination with the agencies of the three spheres of government, as well with religious, financial, academic, and social movements' institutions. The network tries to integrate different types of activities, and seeks to streamline efforts and maximize results during the training process, as well as access families who are benefiting from social programs under its responsibility. Telecenters play an important role as an aggregator link of a network of activities required, so that a significant portion of the Brazilian population can economically and socially integrate. Various institutions have supported the network by way of donating computer equipment for telecenters, thereby enabling significant network expansion.

This approach has shown the manner in which the business community and the public managers rallied for causes that could promote the reduction of inequality of opportunities in Brazil.

In this context of social inclusion, the TIN network can stimulate and strengthen productive activities through:

– access to information on various products, arts, and crafts produced in other regions of the country or abroad;

– training (through computers) in the production and design techniques, techniques of organization of the MSEs, marketing, management, associativism, cooperativism, and fair trade;

– access to government programs and projects oriented toward the promotion of productive activities in agriculture, commerce, industry, tourism, crafts, and services;

– access to information on management, associativism, and cooperativism as forms of organization of productive groups and integration in social economy programs;

– access to market information, primarily, for fair and solidarity trade networks in Brazil and abroad;

– access to information related to tax legislation and the organization of small enterprises, cooperatives, and producer associations;

– use of electronic commerce through the Internet pages which disseminate the local and regional products and services;

– interaction with technical schools, universities, and centers for technological research;

– access to information on credit, microcredit, and e-banking. Public and private banks and organizations which can avail and benefit from microcredit can use of telecenters as platforms to secure these services that interest the entrepreneurs, artisans, and the local and regional community for the implementation of relevant banking interactions, with or without cash.

16.2.3. *Application 2.0 in information and training*

Business and information telecenters are constantly innovating in terms of information and training. An initiative has been undertaken to create a virtual community, which is incorporated into the information and business telecenters' portal. This initiative has different objectives, e.g. facilitate the management of telecenters, promote information and business management, provide subsidies for content development, substantially increase the involvement of teams, promote electronic commerce, allow business meetings, socialize knowledge and stimulate the integration of the network telecenters.

The possibility that a larger number of representatives are likely to attend the virtual meetings would enable this initiative to minimize the travelling expenses of staff and optimize the decision-making process.

Each community can adopt its participation methodology based on the characteristics and interests of its members, and use the available modules (virtual library, meeting rooms, information, discussion forums, surveys, research, etc.). Existing communities are managers of telecenters, TIN-MEC, and managers of telecenters partners, TIN. Communities that are affiliated with the permanent forum of MSEs can also access the portal: overall coordination of the permanent forum on MSEs, thematic information committee, the committee for thematic training and training for entrepreneurship, the thematic committee on trade and international integration, on investment and funds, on legal and bureaucratic streamlining, and in technology and innovation.

But, this is not the only initiative that uses Web 2.0. Ever since its inception, the network has been managing the information and business telecenters' portal by offering basic content for the already installed network, thereby reducing the initial effort that the telecenter has to put in for developing its own content. Over a period of time, through the study of profiles of its users, each telecenter would develop and deliver new content, not just to its community, but also to the entire network.

To determine the needs of each industry, we are regularly consulting public policies, which are likely to assist the selection criteria of information to be included on the portal of business and information telecenters. At present, the selected contents relate to:

– agribusiness: rural administration, agriculture, agro-industry, livestock;

– support for micro and small firms: entrepreneurship, online trade;

– crafts: Brazilian crafts, training and management of the artisan, cooperativism, history, what is panorama craft, why handicrafts are exported, old age insurance, programs, who is the craftsman, craftsmanship techniques;

– science and technology: telescience chain, federal council, support for enterprises, industrial technology infrastructure, recyt[8], (Mercosur Special Meeting on Science and Technology) Information Technology;

– trade: international agreements, agribusiness (farming engaged as a large scale business) trade barriers, international competitiveness, international contracts, customs documents, events, credit instruments, international logistics, first steps, industrial technology, health surveillance, trade and services, service-providing institutions, transport and distribution;

– professional advice: presentation, Ciam (Israelite Center of Multidisciplinary Support – dedicated to the social inclusion of children, adolescents and adults with intellectual deficiencies), regional councils, taxation, legislation, orders and professional advice, projects, training and education;

– education: corporate, general, second cycle of secondary education;

---

8 Recyt: Mercosur special meeting on science and technology, created in June 1992.

– financing: working capital, financing assets, microcredit;

– industry: management and sales support, knowledge of Abimaq (Brazilian Association of Machines and Equipment Industries), legislation, professional guidance, regional, federal, and municipal legislation;

– social programs: children assistance from birth up to 6 years, assistance for people with disabilities, family allowance to help the disadvantaged, fight against sexual exploitation of children and adolescents, zero hunger, total family assistance program, program for the eradication of slave labor, project agenda for the social and human development of the youth.

The portal content is being continually updated. It is also possible to retrieve information by using the hyperbolic navigation tool.

It is worthwhile noting that the TIN network organizes training courses in the form of extension management of telecenters and digital entrepreneurship through distant learning program offered on the Moodle platform. The courses have been specifically developed for training managers of telecenters, with a 64-hour period and are support of specialized tutors. Content is available in five modules, in order to boost entrepreneurial potential and improve management practices of telecenters. At the end of the course, students would receive a certificate from the University of Brasilia. Since 2007, 300 managers of telecenters have been trained, with 85% success rate. In 2009, there were over 338 entries.

16.2.4. *Perpetual beta in management and coordination*

TIN network pays maximum attention to the sustainability of its units. For this, the coordination of the network has established that administrative support should continuously evolve:

– auto sustainability: units put in place must innovate business models to generate sufficient revenue to cover expenses of the telecenter and make new investments;

– collective and multipurpose space: the infrastructure must be sufficient for training activities, IT demonstrations, e-bank, and experiments and services, which will benefit firms in terms of their orientation as company, and for costs reduction, facilitating new acquisitions, increasing their sales, for professional training, improving post-sale, trade automation, allowing them more opportunities, a greater efficiency, new trading models, and digital inclusion;

– accessibility and ergonomic: the proposed telecenter services would plan to install equipments and use technologies that would facilitate access and usage, irrespective of the educational level or physical or mental conditions of its users;

– independence and decentralization: apart from the content and recommendations offered by the institutions responsible for the project and managing the various units, institutions will be able to adapt or create resources, products, and services that are better adapted to the specific requests of their clients;

– forming network: the units would have different communication mechanisms and facilities to improve the integration of the telecenter nodes, so experiences and content sharing are possible;

– diverse content: the proposal of services and products of the units are expandable and are not restricted only to the industrial, commercial, or service sectors. The range of options available to users of the telecenter will be defined in terms of the diversity profile of each application;

– negotiation and articulation: the proposal of products and services of telecenters will not be limited to the available coordination offer. If required, alliances can be made with other institutions for creating new services of interest to the telecenter;

– norms and minimum standard: the regulatory standard of telecenters has established criteria and regulations for the use of computer facilities and network relating to the protection of users privacy. It must be constantly updated according to trends in digital inclusion;

– continuous monitoring: TIN monitoring network is made from a true coordination observatory and will be programmed to ensure the prospective studies, which will enable building and evaluating the evolution of information and communication technology systems:

- to ensure access, collection, and processing of information relevant to the micro and small business,

- to promote the dissemination of knowledge organized by telecenters,

- to develop and keep an updated inventory of TIN network units,

- to study and prepare an estimation bill of the network,

- to promote the development of computer applications which will support the activities of the telecenters,

- to promote and participate in the development of structures, networks, and information systems at national and international level;

– knowledge base: is based on data collected continuously in telecenters, which can provide the requisite information to the MSEs across the country with the development of individual knowledge base. After a consolidation process, we can them move to the creation of the national knowledge base. These knowledge bases will be fundamental tools for studies on the segment, and would formule strategies for local consultation, and national policies or actions.

16.2.5. *Long tail: the real users of the TIN network*

MSEs segment constitute a significant percentage of the national productive sector. They represent about 98% of all business activities in Brazil, employing about 80% of the workforce and account of 42% of the payroll. However, despite its ability to generate jobs as presented, its participation as other economic indicators do not correspond with its real potential. Although they represent about 30% of export companies in Brazil, the export value generated does not exceed 2%.

TIN network operates on the basis of a standard model of telecenter. This model consists of a physical space located in an institution that represents or develops actions oriented toward MSEs. This space has a computer infrastructure and human resources necessary for digital literacy and usage of major Internet resources. Each TIN relies primarily on a manager who manages monitors and assistants. The telecenters have more than eleven networked computers, and are connected to the Internet and other equipments, such as a printer, scanner, fax, telephone, TV, etc. The installation, incorporation, or development of other more specialized units is, therefore, also encouraged. Such units include:

– virtual incubators of projects and companies which develop software, applications, and even content to other units based on a technology independent software;

– cooperatives of MSEs and artisans for large-scale purchasing of parts and selling products of the same brand with high market penetration;

– credit unions that capture and forward credit lines and finance needs for working capital and the investment needs of MSEs affiliated with it;

– specialized telecenters for a determined productive segment (e.g. confections, furniture manufacturers, food manufacturers, etc.) that responds primarily to a local or regional productive arrangement;

– commercial centers which combine the MSEs and provide all the infrastructure, security, and electronic commerce.

The diversity of models is accompanied by a greater diversity of content. The content, (databases, information services, training and vocational courses, etc.), that is generally technological and managerial in nature, takes into consideration the target market with which the served microenterprises are working. The depth, extent, and type of knowledge, as well as the services offered, vary depending on: (1) whether the target market is local, regional, national, or foreign; (2) whether the buyer is the government or a natural or legal person; (3) whether it is in urban (small, medium, and large cities) or rural areas; or (4) whether the competition comes from small, medium, large, or multinational companies (located in Brazil or abroad).

The training of the head of a microenterprise unit and his production team is conducted by using distance and e-learning methodologies and technologies. It provides access to information that takes full advantage of tacit knowledge and cultural background of an individual and community, thereby allowing him to update his popular technological culture in relation to the technical advancement of production methods (gathering, drying, storage, filtering, storage, etc.) and quality control.

To illustrate the type of information, software, or services available in a TIN in terms of content diversity, the company manager can access knowledge from the indicators based: (1) on the usage of project technology and product development; (2) on the manufacturing, preservation, packaging and transportation process of products and materials; (3) on the organization, planning, and production control; (4) on adaptive management for small business (craft or semi-industrial production) of low complexity that has a technological sophistication oriented toward the use of locally or regionally available resources and raw materials; (5) on machinery and tools adapted to tranches of microcredit funding and the investment capacity of the company manager or of his cooperative; and (6) on divulging the basic concepts of standardization, certification, as well as the culture of the use of standards and implementation of quality controls suitable for MSEs, with or without access to technical or technological services.

16.3. Social and economic impact of the TIN network: some indicators

The data collected indicate that 614 staff was trained (managers, educators, multipliers telecenters), more than 110,000 users have accessed the TIN portal, 1,198 new units were put in place, and about 11,980 computers have been distributed. We have observed that we would need more than 6 million dollars to build the same number of units equipped with new computers (valued at $1,400, the estimated price of each computer in the context of the "Computers for Everyone" program of December 31, 2006). It should be noted here that in 2006 the government had allocated a budget worth only 1 million dollar to the business and information telecenters project.

Currently, the project has installed more than 1,500 units (including the associated units – existing networks that have joined the project in an effort to digitally include MSEs). This figure represents more than 20,000 computers or more than 20,000 gateways to information, which are fundamental for the development of the country. Achieving this threshold could mean that over 30 million dollars of the budget has been utilized, if we assume by this calculation that spending has only been on equipment purchase (in the case where the project would use only new computers in the installation of telecenters).

In addition, this government action creates public usage for technological waste of some macroinstitutions of the national economy. This way, used computers have reduced the digital divide in Brazil.

Business and information telecenters are present in all units of the federation, with more than one hundred partner institutions which support their implementation, apart from the institutions that house the telecenters.

However, the main indicator of competence is the recognition of the community where the telecenter network is operating. In this scenario, during 2007, the contents of the portal had enabled telecenters to get the World Summit Award (WSA) in the category of   e-inclusion (digital inclusion). The WSA award, a global initiative, is based on criteria relating to content and digital creativity to select and promote the best content and applications on virtual Internet. There were more than 650 indications from 160 countries for the WSA price. The grand jury chose winners for the top five products in each of its eight categories. This award is sponsored by the United Nations Industrial Development Organization (UNIDO), UNESCO, and the Internet Society. It is supported by the Brazilian Ministry of Science and Technology.

Internationally, the TIN network has also been recognized by UNIDO in 2006, where such recognition would serve as a model to be transferred to other countries.

In addition, during 2006, on the national scene, the distinction received from B2B Magazine on E-consulting, wherein the TIN network had received the "quality standard in B2B" price in the "digital corporate responsibility in the public sector" category, deserves special attention. The judgement is based on the ability to increase the competitiveness and innovative potential of Brazilian organizations.

16.3.1. *Telecenters in remote areas*

Telecenters, with computers donated by *Schering do Brasil*, was set up in the Rondonia state of the Amazonia region. The telecenters were primarily used to enable the connection of four exporting units in Rondônia, with support from the association of MSEs in the state. In this way, these exporting units have been able to strengthen their communication (including correspondence courses) features with the central office located in the capital, Porto Velho, and could respond to requests even from abroad through the Voice-over IP Technology.

16.3.2. *Telecenters in electronic commerce*

In the Piaui state, the first telecenter of the TIN network has enabled small local enterprises to expand their business through electronic commerce, as well as the intensive and creative usage of email as a tool for communication and business marketing.

In the Federal District, the Ceilândia telecenter (near Brasilia) has already been empowered to respond to requests for online purchase. Dozens of managers of micro and small businesses have been trained in the usage of the electronic commerce tools, as well as the Internet access. Many websites have also been created and published, enabling a significant reduction of the digital divide of the micro business in the realm of the Federal District.

16.4. Telecenters and competitive intelligence: the future of Innovation 2.0

The 2.0 concept, a new concept of Tim O'Reilly [ORE 05] which is oriented toward virtual environment, right from its inception, has inspired competitive intelligence actors to intensively use the understanding of the phenomenon in everyday life of the company. With additional adjacent ontologies, expression has been expanded to explain the reason for the success of several entrepreneurial endeavors, especially in the field of technology and innovation.

In Brazil, a concrete application of the concept of 2.0 is in the network of business and information telecenters (TIN) – the Brazilian government program whose results are evident in the digital inclusion of micro and small companies. This is a national action of competitive and economic intelligence which seeks to shape the competitiveness of this segment in Brazil.

The idea was to think of innovation in different ways: in the network structure which has allowed the usage of concepts, such as the triple helix in government – academia partnership and the private sector, which is an essential development synergy. To establish this synergy, it was necessary to adopt an organizational model which was different from that of the bureaucratic organization, one that Mintzberg and McHugh have termed "adocratique[9]".

---

9 Its name has originated from the Latin term *ad hoc*, i.e. created for a particular purpose, transient, eventual. Adocratique is an organization formed for a limited time and its fluctuation depends on the needs of the moment. These are volatile organizations, whose main features include the organization of teamwork, bonds structured by teams whose main task is to coordinate their own work.

To lend credibility to the project and enable it to be sustainable, it was necessary to establish an international interconnectivity which could only be conceived through a deep understanding of what constitutes "cultural dimensions", as described by Geert Hofstede[10].

Geert Hofstede visualized an innovation that was oriented and focussed on capacity building, thereby enabling the social and economic development of firms, and influencing social changes in their communities.

Future expansions of the TIN network include innovation from the perspective of direct development of competitive intelligence: the identification of information needs in response to specific problems, and environmental scanning through sectorial corporate education. In this context, the 2.0 concept is renewed permanently through the Semantic Web, joint partnerships, social concerns expressed in learning throughout one's life by feeding a desire for continuous improvement, and finally, seeking to fill the niches of the long tail.

16.5. Bibliography

[AMA 97] AMARTYA S., "Human capital and human capabilities", *World Development*, vol. 25, no. 12, p. 1959–1961, Elsevier Science Publishers, Quebec, Canada, 1997.

[AMA 99] AMARTYA S., *Development as Freedom*, Oxford University Press, Oxford, 1999.

[AMA 00] AMARTYA S., *Desenvolvimento com liberdade*, Companhia das Letras, São Paulo, 2000.

[AND 06] ANDERSON C., *The Long Tail: Why the Future of Business is Selling Less of More*, Hyperion, New York, 2006.

[ARR 06] ARRUDA M., Inclusão digital das empresas brasileiras, Ministério do Desen-volvimento, Indústria e Comércio Exterior, Brasilia, Brazil, 2006.

[BRA 04] Brazilian Ministry of Development, Industry and Foreign Trade, *Integração de iniciativas interinstitucionais ao Fome Zero: estratégias de aproximação do sistema nacional de ciência e tecnologia*, MDIC, Brasilia, Brazil, 2004.

---

10 Scientific study of management in diverse cultural contexts.

[BRA 05] Brazilian Ministry of Development, Industry and Foreign Trade, *Regimento interno dos Telecentros de Informação e Negócios*, MDIC/Secretaria de Tecnologia Industrial, Brasilia, Brazil, 2005.

[BRA 06a] Brazilian Ministry of Development, Industry and Foreign Trade, *Manual do Gestor de Telecentros de Informação e Negócios: orientações gerais*, 1e édition, MDIC, Brasilia, Brazil, 2006.

[BRA 06b] Brazilian Ministry of Development, Industry and Foreign Trade, *Telecentros de Informação e Negócios: o desafio da inclusão digital da microempresa e empresa de pequeno porte*, 1e édition, MDIC, Brasilia, Brazil, 2006.

[BRA 08] Brazilian Ministry of Development, Industry and Foreign Trade, *Integração e consolidação das iniciativas de inclusão digital do MDIC e do Governo do Estado de Minas Gerais*, MDIC/Secretaria de Tecnologia Industrial, Brasília, 2008.

[LEY 98] LEYDESDORFF L., ETZKOWITZ H., "The Triple Helix as a model for innovation studies", *Science and Public Policy*, vol. 25, no. 3, p. 195-203, 1998.

[NOR 94] NORTH D., *Institutions, Institutional Change and Economic Performance*, Cambridge University Press, Cambridge, 1994.

[ORE 05] O'REILLY T., O que é Web 2.0: padrões de design e modelos de negócios para a nova geração de software, 2005, available at http://www.oreilly.com.

[QUO 08] QUONIAM L., *Discours du Godfather Docteur Honoris Causa à José Rincon Ferreira*, Fernando Pessoa University, Porto, 2008, available at http://www.educor.gov.br/noticias/drlucquoniam.htm.

[RAM 05] RAMOS P.A.B., *Tecnologias apropriadas e inclusão digital: criando oportunidades de emprego e renda na medida certa*, MDIC, Brasilia, Brazil, 2005.

[STI 02] STIGLITZ J., *Globalization and its Discontents*, W.W. Norton & Co., New York, London, 2002.

[TAR 07] TARAPANOFF K., *Histórico dos Telecentros de Informação e Negócios*, MDIC, Brasília, Ministério do Desenvolvimento, Indústria e Comércio Exterior (Ministry of Industry and Foreign Trade Development), Brasilia, DF, Brazil, 2007.

[TOU 05] TOURAINE A., *Un nouveau paradigme: pour comprendre le monde d'aujourd'hui*, Fayard, Paris, 2005.

[YOL 04] YOLIN J.M., *Internet et entreprise: mirages et opportunités*, Ministère de l'Economie, des Finances et de l'Industrie, Paris, 2004.

Chapter 17

University: Catalyst for the Implementation of Competitive Intelligence 2.0 in Africa (Case Study of Nigeria)

17.1. Introduction

African countries such as Nigeria, have not incorporated Competitive Intelligence (CI) as part of their public policy. On the other hand, other countries such as Brazil and France have been able to identify the virtuous circle generated by the triple helix concept, which resides in the networking of states, industries, and universities. Therefore, the CI concept is being introduced through training and research, mainly in universities through inter-university networks established with Western countries.

In fact, competitiveness is not the main concern or the source of motivation for development in Nigeria. The Nigerian interest and development policy focus on the means, methods, and technologies for more effective governance of the country in all socioeconomic aspects. Therefore, academic links seem to be relevant methods as the first step in long-term strategic implementation. We will, therefore, present the genesis of the introduction of Economic Intelligence (EI) in Nigeria, and then we will consider the interest in setting up international projects and highlight the development opportunities offered by the establishment of EI in Nigeria.

Chapter written by Amos DAVID.

17.2. Genesis of the introduction of EI in Nigeria

Chronologically, the introduction of EI in Nigeria commenced in 2002 with the signing of a collaboration framework agreement between the University of Nancy 2, France, and University of Ibadan, Nigeria. The latter is the country's first university, established in 1948; its areas of interests include teaching and research centers. The framework agreement is the result of the contacts established with the Nigerian universities through the EDUFRANCE agency mission in collaboration with the French Ministry of Foreign Affairs and the Department of Cooperation and Cultural Action (SCAC)[1] of the French Embassy in Nigeria. In fact, the mission which had reached out to four Nigerian universities between 2000 and 2001 was meant to present French higher education training. The aim was to present French excellence in terms of higher education and the French policy for teaching and research. It should be noted that Nigeria is an English speaking country.

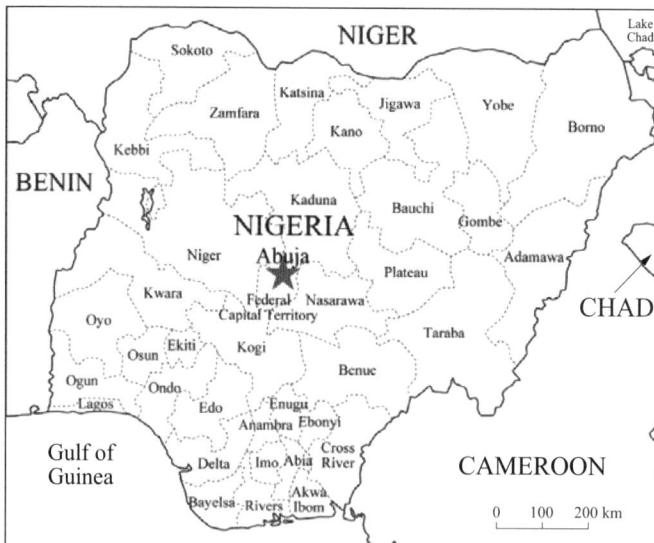

Figure 17.1. *Map of Nigeria*

However, Nigeria is surrounded by French-speaking countries: in the East by Cameroon, North-East by Chad, North by Niger, and West by Benin. Although the students are aware of the training opportunities (degrees and

---

1 *Le Service de Coopération et d'Actions Culturelles.*

costs) in English-speaking countries, they are completely ignorant of the programs offered in France. The implementation of academic links here implies a real political influence for future cultural and trade relationships.

17.2.1. *Implementation of a collaborative strategy*

The initial objective of the collaboration was to facilitate mutual understanding of the cultures and the education policies of Nigeria and France. Thus, the SCAC of the French Embassy in Nigeria supported the organization with a conference on the theme "Information and Communication Technology Applied to Economic Intelligence" in July 2002, at the University of Ibadan; the proceedings were published by INRIA editions [DAV 02].

Another main focus is on helping in the development of human resources in Nigerian universities, through training of lecturers and participation in students training in Nigeria by French lecturers, and also through funding for equipment.

In 2002, although information and communication technologies had started in Nigeria, the institutions (universities and the SCAC) defined common objectives based on mutual recognition and trust. In fact, there was no precedent action in Nigeria in terms of cooperative approach in the domain. The mode of cooperation in implementation was largely *ad hoc* and was based on fixed-term projects. The mission was organized around a long-term approach based on collaboration between universities at the national and international levels.

Four Nigerian universities were encouraged to collaborate in organizing the conference of 2002 and then a workshop in 2003. The conference was the first concrete collaboration action between four universities in South-Western Nigeria (University of Ibadan, Obafemi Awolowo University, University of Lagos, and Lagos State University). In this manner, the culture of information sharing and resources pooling culture was introduced in the universities. The adoption of this culture has facilitated the success of all the other projects implemented in this collaboration framework. As a result of the success of the conference and the encouragement of SCAC of the French Embassy in Nigeria, the following projects were implemented:

– 2003: a workshop was held at the University of Ibadan on "Design and Development of Internet-Based Information Systems";

– 2004: a summer school was organized on "Advanced Topics in Databases", which is one of the courses offered at the Master's program of the computer science department at the University of Ibadan. The other three universities participated in the training;

– 2005, 2006: summer school was also organized;

– 2007: to respond adequately to the increasing number of participants of the summer school from the four universities, two sessions were held. These sessions included a session for the University of Ibadan and Obafemi Awolowo University, and the other session for the University of Lagos and the Lagos State University;

– 2008: two training sessions were also held.

At the international level, international collaboration was organized around scientific visits of Nigerian lecturers to France, and scholarships were sought for students at professional master and doctorate levels. The Lorraine region in France had provided funding for one month scientific stay for three Nigerian lecturers in 2003 as part of the collaboration.

17.2.2. *Building on existing structures*

For the implementation of projects under collaboration, instead of creating new structures, the preference was to rely on the existing structures – universities, training departments, and their staff – and to give them support.

17.2.2.1. *University structures*

In 2002, Nigeria had more than 40 universities. Many of these universities have a standard of excellence which is internationally recognized. Academic structures were, therefore, relevant networks for the implementation of collaboration projects.

An advantage of this option is the availability of an already proven organization and a system of control and empowerment of a level and quality sufficient for training opportunities. It is relatively easy to propose new training programs in an existing faculty or in a department within a faculty. It is also quite easy to adapt existing training to integrate new educational content. This advantage allowed us, starting in 2004, to incorporate a training module on the development of Internet-based information system in an existing master program.

Another advantage is the open policy which is adopted by the universities for international collaboration. It was, therefore, quite easy to convince the management board of the universities of the importance of the type of collaboration proposed. In summary, the following projects were realized, based on the existing structures of the university:

– 2002: University of Nancy 2 hosted two PhD students from Nigeria, including one from the University of Ibadan, and the other from the Obafemi Awolowo University. Both theses were defended in 2007 and the two doctors have since then returned to their respective universities to join the teaching staff;

– 2007: three faculty members of the University of Nancy 2 conducted a research visit of three weeks to the four Nigerian universities involved in the collaboration;

– 2007: four lecturers from Nigeria, one from each university involved in the collaboration, conducted a one month study visit to France. They were hosted by the research team SITE-LORIA;

– 2007: four PhD scholarships were awarded to doctoral students (two from the University of Ibadan, one from the University of Lagos, and one from Lagos State University. Obafemi Awolowo University did not propose a candidate). These scholarship grants were for co-supervised PhDs. Each student spent five months in France and seven months in Nigerian every a year for a period of 3 years;

– 2008: four Nigerian lecturers, one from each university, were involved in the collaboration, and conducted a one month study visit to France. They were hosted by the research team SITE-LORIA.

However, some difficulties were encountered and ultimately overcome. A specific example concerned the identification and availability of a contact person to monitor and ensure the success of the projects. Indeed, most of the lecturers are already engaged in different administrative responsibilities.

17.2.2.2. *National ministerial bodies*

The Nigerian government's national policy which gives priority to ICT development, both in terms of infrastructure and training, has also greatly facilitated the implementation of the projects.

17.2.2.3. *Agencies and international policies*

The start of the collaboration in Nigerian universities also corresponds to the commencement of a policy for international collaborations in French universities. Therefore, in 2002, Nigerian students were granted scholarships for the professional Master and doctoral study programs. This policy is currently being maintained through encouragement of collaboration in PhD studies through co-supervised theses. For example, four co-supervised PhDs are being funded by the French Ministry of Foreign Affairs for Nigerian PhD students from the universities involved in the collaboration. It was through these PhDs that training in EI was concretely introduced. In addition, in 2007, two lecturers from the University of Ibadan were employed as Associate Professors by University of Nancy 2, each for a period of one month to give lectures in the Master's program specialized on EI, and to conduct research in the research team SITE-LORIA.

The French policy in encouraging international collaboration also covers facilitating double-degree Master programs. Therefore, a proposition of masters in EI was formulated and authorized in France. This collaboration has enabled an investment in a wider collaboration at the African level, particularly through the SIST project which is presented in the next section.

17.3. Participation in international projects

The collaboration has enabled the development of a Franco-African project – SIST (*Système d'Information Scientifique et Technique*[2]) – comprising 13 African countries and France[3]. It includes the development and deployment of an information system platform for managing information resources and searching on local and external databases.

---

2 Scientific and technical information system.
3 http://www.sist-sciencesdev.net/.

17.3.1. *Toward the creation of a research institute in SIS-EI*

The implementation of EI is limited to the training framework – graduate, post-graduate and research training. This experience could be extended by creating a research institute which would focus on the development of strategic information and business intelligence systems. Three main reasons for this orientation are:

– a research institute should facilitate knowledge transfer between researchers and businesses or, in a broader sense, socioeconomic organizations;

– a research institute is better organized to establish research and development programs, thus facilitating collaboration between a socioeconomic organization and the research world. This is in line with the first reason given above;

– research is carried out at the departmental level in universities and is rarely structured into laboratories. A research institute has a structure that is similar to the research laboratories in Europe or the United States. The establishment of a research institute would facilitate direct collaboration between laboratories in Europe or the United States and, most importantly, would allow researchers in other Nigerian universities to collaborate together on shared themes.

17.3.2. *Support of the Nigerian diasporas in France*

In terms of political and financial support, the Nigerian diasporas in France were mobilized during a conference on "ICT Development in Nigeria: impact on sustainable development at state and local government levels", held in July 23–24, 2009[4]. This conference aimed to develop collaborations and partnerships for the transition of states and local governments of Nigeria into an economy based on information and communication technology (ICT). In fact, Nigeria has experienced significant progress in recent years, throughout the country, in terms of development of information and communication infrastructure, thereby reducing the digital divide. Efforts are needed from public and private sectors, from local communities and diasporas organizations and institutions, to accelerate the process to a rhythm proportional to ICT industry dynamics.

---

4 http://nidoefrance.org/Commune.aspx (online on 20/07/2009).

The recent fluctuations in oil prices have revealed the need to develop sources of income which are less dependent on hydrocarbons. In particular, the state and local governments should intensify their ongoing efforts to improve the process of collecting revenue and diversifying the sources of income. Opportunities exist in the global export market for goods and services. The consequences of globalization are forcing states to create conditions for companies that operate locally, so that these companies are not only able to withstand competition in the local market, but are well-armed against the international competition. Macroeconomic policies are not sufficient to promote competitiveness of firms operating in a nation.

To improve the position of domestic firms in the global market, both government policies and good microeconomic strategies are required. In the absence of coordinated contribution of both sectors in well-defined frameworks, growth will be compromised not just at the federal level, but also at the state and local government levels. The role of ICT and EI strategy as public policy proves to be a key components of economic development in emerging countries. It provides a solid foundation for sustainable economic growth, job creation, poverty reduction, and income generation for governments. The various state and local governments' partners and stakeholders are concerned about:

– federal, state, and local e-governance authorities, the development of trade and industry, efficient automation of revenue collection and administration of national security, information, and control systems;

– national companies who are targeting the global market and are required to promote their products and services for export;

– foreign companies and investors who are already operating in Nigeria or those who desire to exploit the enormous market potential currently existing in the country;

– ICT professionals in the country and the diasporas: for the exchange of knowledge and know-how which are essential for effective technology transfer;

– international organizations concerned with the development of emerging countries;

– other Nigerians from various diaspora organizations who are interested in national development.

17.4. Economic intelligence: a developmental perspective for Nigeria

The interest of the academic approach in implementing an EI policy is seen in its bottom-up approach, which is fully in line with the 2.0 concept. Indeed, by training and educating the academics on the 2.0 concept, the approach would be proposed to the industrial world. The information sharing and open approach to collaboration culture, along with networking (which excludes the rigidities that characterize a pyramidal organization) must be shared by all the employees in a company.

17.4.1. *Bridging the digital divide, a development challenge*

Implementing a vision of 2.0 and a realistic EI strategy, not only implies the possibility of networking actors in organizations and networking territories, but also incorporating a "thought from outside", i.e. a permanent self-questioning feature.

In this sense, strategies for expanding access to ICTs are in fact part of the development policies to a greater extent. It entails influencing the social architectures and overturning interactions. ICT, through transfers of skills and partnerships, enables the organization to catch up with heavy delays. Henri Dou [DOU 05] has particularly shown the impact of EI approaches on developing countries, including Thailand and the Philippines. The availability of patent databases allows the democratization of industrial information and opening of new markets. Coconuts, dates, or palm oil, therefore, assert themselves as riches hitherto despised, and yet promise sure and unexpected economic development.

17.4.2. *Toward a redefinition of competitive intelligence?*

The term CI connotes two possible interpretations depending on the role of the adjective "competitive". In a first sense of the term, CI refers to a set of concepts, techniques, and methods designed with the objective of enabling the company to have competitive advantage over its competitors[5]. Thus, "CI is the process of monitoring the competitive environment and analyzing the findings in the context of internal issues, for the purpose of decision support. CI enables senior managers in companies of all sizes to make more-informed

---

5 http://fr.wikipedia.org/wiki/Intelligence_comp%C3%A9titive (01/07/2009).

decisions about everything, from marketing, R&D, and investing tactics to long-term business strategies. Effective CI is a continuous process involving the legal and ethical collection of information, analysis that does not avoid unwelcome conclusions, and controlled dissemination of actionable intelligence to decision makers[6]". In the first sense, the main motivation is the competitiveness in a competitive environment.

The second meaning of the term would be the translation of EI[7]. As presented on the website of the senior EI officer in France (HRIE), "... they understood that effective organization depends on information sharing, networking, exchange of ideas and knowledge, and dissemination of synthesis..." "...EI is a form of management, a policy of controlling information, can be applied to all sectors, and not just those that are deemed 'strategic'....[8]"

In this second sense, EI is not limited to competitive contexts, but covers all areas of decision problem resolution, not necessarily integrating itself into a market policy [DAV 03, DAV 05]. Emphasis is also placed on the need for protection and enhancement of tangible and intangible heritage, such as know-how, without necessarily demonstrating liberalism and protectionism linked to the beginnings of EI in the 1980s. Thus, the adjective "competitive" is not only related to the competitive environment, but also to the production of intelligence that is competitive [DAV 09]. In this case, CI acquires a dimension of territorial development and includes it in humanistic values and intercultural exchange.

17.5. Bibliography

[AUB 04] AUBERT J.E., Promoting innovation in developing countries: A conceptual framework, World Bank Institute, July 2004, available at http://siteresources.worldbank.org/KFDLP/Resources/0-3097AubertPaper%5B%5D.pdf.

[DAV 02] DAVID A., OSOFISAN A., Information and Communication Technologies Applied to Economic Intelligence, Edition INRIA, Ibadan, Nigeria, July 2002.

[DAV 03] DAVID A., IE: Recherches et Applications, Edition INRIA, Nancy, April 2003.

---

6 http://www.scip.org/content.cfm?itemnumber=2214&navItemNumber=492, 01/06/2009.
7 http://www.intelligence-economique.gouv.fr/rubrique.php3?id_rubrique=40, 01/06/2009.
8 http://www.intelligence-economique.gouv.fr/rubrique.php3?id_rubrique=8, 01/06/2009.

[DAV 05] DAVID A., *Organisation des connaissances dans les systèmes d'informations orientés utilisation: contexte de veille et d'IE*, Presses Universitaires de Nancy, Nancy, 2005.

[DAV 09] DAVID A., "Relevant information in economic intelligence", in *Information Science*, ISTE Ltd., London, John Wiley & Sons, New York, 2009.

[DOU 04] DOU H., MANULLANG D.S., "Competitive intelligence and regional development within the framework of Indonesian provincial autonomy", *Education for Information*, no. 22, June 2004.

[DOU 05] DOU H., DOU J.M. JR, MANULLANG D.S., "The magic triangle. How to develop and apply competitive intelligence in developing countries", *Journées sur l'Information Elaborée*, Ile Rousse, France, June 2005.

[LEY 98] LEYDESDORFF L., ETZKOWITZ H., "The triple helix as a model for innovation studies (conference report)", *Science & Public Policy*, vol. 25, no. 3, p. 195-203, 1998.

List of Authors

Lillian ALVARES
Faculty of Information Science
University of Brasilia
Brazil

Wanise BARROSO
CTM/Farmarguinhos Department
Brazil

Sébastien BRUYÈRE
Paragraphe Laboratory
University of Paris 8
France

Bruno Filipe CARVALHO SOARES
University Fernando Pessoa
Porto
Portugal

João CASQUEIRA CARDOSO
University Fernando Pessoa
Porto
Portugal

Serge CHAUDY
University of the South Toulon-Var
France

Amos DAVID
Equipe SITE
LORIA
France

Christophe DESCHAMPS
University of Poitiers
and EISTI Cergy-Pontoise
France

Fabrizio DOLFI
NicOx
Sophia Antipolis
France

Henri DOU
Beijing University
China

Patricia DUPIN
Intelligentsia CI
Brazil

Brigitte GAY
Group ESCT Business School
Toulouse
France

Lucia GRANGET
Institut Ingémédia
University of the South Toulon-Var
France

Zhouying JIN
Chinese Academy of Social Sciences (CASS)
Beijing
China

Philippe KISLIN
INDEX
Paragraphe Laboratory
University of Paris 8
Equipe SITE
LORIA
Nancy
France

Arnaud LUCIEN
University of the South Toulon-Var
France

Fabrice MAULEON
Groupe ESCEM
Tours
France

Rosana PAULUCI
Lab4U
Research Laboratory in Science of Information & Communication
France

Jean-Dominique PIERRET
Salon-de-Provence
France

Joachim QUEYRAS
DATAR
Paris
France

Luc QUONIAM
Paragraphe Laboratory
University of Paris 8
France

José RINCON FERREIRA
JFR Prospect
Manaus
Brazil

Miguel ROMBERT TRIGO
University Fernando Pessoa
Porto
Portugal

Alice Maria SALGADO GONÇALVES
Consultant in Processing and Analyzing Statistical Data
Porto
Portugal

Kira TARAPANOFF
Faculty of Information Science
University of Brasilia
Brazil

Index